Space, Curriculum, and Learning

A Volume in
International Perspectives on Curriculum
Series Editor: David Scott

Space, Curriculum, and Learning

Edited by

Richard Edwards and Robin Usher

INFORMATION AGE PUBLISHING

80 Mason Street • Greenwich, Connecticut 06830 • www.infoagepub.com

Library of Congress Cataloging-in-Publication Data

Space, curriculum, and learning / edited by Richard Edwards and Robin Usher.
p. cm. – (International perspectives on curriculum)
Includes bibliographical references.
ISBN 1-59311-092-8 (pbk.) – ISBN 1-59311-093-6 (hardcover)
1. School environment–Social aspects–Cross-cultural studies. 2. Distance education–Social aspects–Cross-cultural studies. 3. Curriculum planning–Cross-cultural studies. 4. Space and time. I. Edwards, Richard, 1956 July 2- II. Usher, Robin, 1944- III. Series.
LC210.S63 2003
375–dc22

2003021884

Printed in the United States of America

CONTENTS

LIST OF ILLUSTRATIONS

TABLES

FIGURES

FOREWORD

The purpose of the series *International Perspectives on Curriculum* is to provide scholarly and authoritative debate about current curriculum issues. The series includes overviews of research in this area, examination of theoretical models and principles, discussion of the work of key curriculum theorists, and the reporting of new empirical research. Contributors to the various volumes in the series are not asked to provide definitive answers to questions that theorists and practitioners working in this field are asking. What they have been asked to do is to critically assess ways of thinking, influential models and current policy initiatives that relate to the curriculum.

The curriculum is defined in its widest sense, and it refers to programmes of teaching and learning which take place in formal and informal settings. Examples of formal settings are schools, colleges and universities. A curriculum may refer to a system, as in a national curriculum, an institution, as in the school curriculum, or even to an individual school, as in the school geography curriculum. It may also refer to programmes in less formal settings. The four dimensions of curriculum are: aims and objectives, content or subject matter, methods or procedures, and evaluation or assessment. The first refers to the reasons for including specific items in the curriculum and excluding others. The second refers to the knowledge, skills or dispositions implicit in the choice of items, and the way that they are arranged. Objectives may be understood as broad general justifications for including particular items and particular pedagogical processes in the curriculum; or as clearly defined and closely delineated outcomes or behaviours; or as a set of appropriate procedures or experiences. The third

Spoace, Curriculum, and Learning, pages xi–xii

dimension is methods or procedures and this refers to pedagogy and is determined by choices made about the first two dimensions. The fourth dimension is assessment or evaluation and this refers to the means for determining whether the curriculum has been successfully implemented. A range of issues have been surfaced and debated in relation to these four dimensions.

The series focuses on these issues and debates. The first volume examined the relationship between Curriculum, Pedagogy and Assessment. The second volume focused on Citizenship Education. This volume is a collection of succinct, highly readable and focused essays on the theme of spatiality and its significance for curriculum and learning. Each volume will take a cross-sector and comparative approach. This series is timely as administrators and policy-makers in different parts of the world have taken an increased interest in education, and as moves to centralize curriculum provision have gathered pace. This has in some cases driven a wedge between curriculum theory and curriculum practice, as policy-makers have developed and implemented proposals without referring to academic debates about these issues. It therefore seems to be an important task to reassert the need to discuss and debate the curriculum in a critical manner before implementation occurs. This series will attempt this difficult, but much needed, task.

—David Scott, *Series Editor*
Lincoln University, UK

CHAPTER 1

PUTTING SPACE BACK ON THE MAP OF LEARNING

Richard Edwards and Robin Usher

In recent years there has been increasing interest in issues of space and spatiality in the social sciences and humanities generally, if less so in the study of education. At one level, this relative lack of interest is surprising given the importance of space and time in the organization of teaching, learning and research. For instance, the timetable and project timeline are central to the organization of learning and knowledge production whether in schools, colleges or universities. Through these technologies people and objects are distributed to certain places for certain lengths of time in a systematic way. Classrooms, workshops and laboratories have different spatial layouts, which support certain forms of interaction and communication. When we add to this, the increasing distances across which knowledge, understanding and competence are being distributed through the use of information and communications technologies, the fact that issues of space and the spatial ordering of people and places have not been taken up seems more than an oversight. This relative lack of interest in space becomes even more surprising when one considers the extensive use of spatial metaphors in the discussion of education and pedagogy. For instance, the notions of open, distance and distributed learning and student-centeredness, border crossing, and communities of practice all have a spatial dimension to them. Notions of a spiral curriculum act as a spatial imaginary (Lawton, 1984). Indeed some metaphors, such as flexibility—

Space, Curriculum, and Learning, pages 1–11

e.g., flexible learning—seem to be suggestive of the possibility that all constraints of space and time can be conquered in the provision of learning opportunities throughout life.

The importance of space and the spatial is at one level obvious therefore, at least once it is pointed out. It has become more pronounced with the range of educational reforms that have taken place around the globe in recent years. These have provided changed spaces for learning and knowledge production, which in turn, have altered the ways in which people interact in learning environments, the constituencies of groups of learners and the forms of learning and knowledge that occur. Take for instance the notion of lifelong learning, which has become a significant part of academic and policy debate in a range of countries in recent years. Lifelong learning provides a temporal framing for learning from the cradle to the grave. However, it also involves spatial re-orderings of learning that have been largely overlooked. For instance, different spaces for learning emerge in the workplace and community, or are given greater significance in these altered states. There is thus a reconfiguration of learning and knowledge spaces through different pedagogic practices, in particular those associated with information and communications technology. What then are the implications for curriculum, learning and identity of these changed spatial relations?

Despite some work on the use of space in school classrooms (e.g., Comber & Wall, 2001), students experience in higher education (Nespor, 1994) and the spatial representations of educational change (Paulston, 2000), there has been little sustained discussion of the spatial dimensions of education. Much of the discussion of educational changes has remained at the level of technical implementation, with few attempts to provide a wider framing, which explicitly highlights the spatial ordering of curriculum and learning. Thus Peters (1996, p. 106) has pleaded for "educational theorists to take seriously questions of space" and to also to develop a politics of space. In this book, we take up this plea.

What we and the contributors to this book wish to do therefore is explore the significance of spatiality for our understanding of curriculum and learning in a range of different contexts, embracing schools, colleges and universities. There are two aspects to our interest in the spatial. The first is what can broadly be termed sociology of space, examining the spatial orderings in specific pedagogic practices and the forms of knowledge, learning and identity they include and exclude. Second we wish to explore the spatial and spatializing metaphors in the discourses of education. In what ways do these order and reorder pedagogic practices? For instance, the notion of globalisation may at one level be taken to be a literal description of trends towards greater inter-connectedness in the world and the experience of the compression of space-time (Harvey, 1991; Tomlinson,

2000). However, it can also be said to be a spatializing metaphor, meanings of which can be drawn upon in, for instance, the development of curriculum about the contemporary world, or in developing distributed learning opportunities for a potential global constituency.

In both conceptions of space and spatiality, the notion of order draws upon the work of Foucault (1979) wherein power and the organization of bodies in space are interactively linked. Thus to talk about space in relation to curriculum, learning and knowledge production is also to talk about how power is distributed and exercised. Many contributors to this text draw upon the work of Foucault or work derived from him, as he has been very important in identifying the centrality of the spatial to social practices, such as education. At the same time for Foucault:

> the present epoch will perhaps be above all the epoch of space. We are in the epoch of simultaneity: we are in the epoch of juxtaposition, the epoch of the near and far, of the side-by-side, of the dispersed. We are at a moment, I believe, when our experience of the world is less that of a long life developing through time than that of a network that connects points and intersects with its own skein (1986).

As we shall see in the contributions to this text, the metaphor of the network is itself powerful—one, that has been taken up in trying to think through the significance of educational spaces.

The chapters herein focus on and bring out for explicit debate and discussion some of the spatial aspects at play in education, and the ways in which the spatial ordering results in particular forms of learning and identity. In focusing on the spatial this book can be considered, with due acknowledgement to C. Wright Mills (1983), as representing a range of cartographical imaginations. In 1959, Mills published *The Sociological Imagination* in which he argued that men (sic) not only need information and reasoning skills, but also a quality that enables them "to achieve lucid summations of what is going on in the world and of what may be happening within themselves" (Wright Mills, 1983, p. 11). This sociological imagination enables people to shift their perspective and to relate the impersonal and remote to the intimate and personal. It is through this imagination that what is happening in the world and in oneself is grasped and the contribution to progress established. This book was hugely influential and focused on the relationship between history, biography and society. In other words, it privileged the temporal dimension, with the sociological imagination in effect constituted as a gendered temporal imagination.

What we are exploring in this text, however, is that with increased attention being given to the spatial, there is now a need to develop cartographical imaginations. These are about geography, inter-connectedness and history (Gregory, 1994). They are plural and relational, and taking care not to fall into another unhelpful binary they embrace time as well as space, with each entailing the other, for example, space-time. It is through cartographical imaginations—investigating the spatial orderings of learning—which we begin to embrace differing understandings of the practices in which we are involved and the ways in which generative spaces for learning and knowledge production are actively constructed.

It is within this framing that we asked colleagues from a number of countries around the English language world to outline proposals for a chapter for this text. They were asked to do this because of the prior work they had done in this area. This work has been published in a range of different formats in a number of places—academic journals, books, and conferences, online. Our aim in putting this text together is to attempt to bring this work together, to provide a focal point—a spatial metaphor itself—through which to examine empirical, theoretical and methodological issues around space, curriculum and learning as a basis for exploring the current state of knowledge and potential directions for the future. A focal point implies a shared interest in spatial issues, but the perspectives offered are diverse, reflecting the different theoretical and methodological positions adopted by different writers.

In the process of putting this text together, we have ourselves as editors become involved in a particular type of spatial practice, associated with editing written texts. We have become the spiders at the centre of a web of communication, largely through the use of information and communications technologies, in the bringing together of the contributions to this text. In other words, we have experienced the networking referred to above in the quote by Foucault and, in the process, constituted a network of those interested in questions of spatiality and education. Our task in this introduction, as is the case for all edited texts, is to provide a network of ideas and understanding from the different contributions, to make connections and links and point to the currently unconnected. As readers, you will make your own connections and webs of meaning. We can only point to our own as editors.

As we have said, space is an obvious focus for attention. Schools, colleges and universities are built environments. They are materially structured in particular ways. Closed classroom, open plan areas, the tiered lecture theatre, the laboratory, the learning resource centre are all designed with particular purposes. As Jenni Karlsson points out so compellingly, in her chapter on what might be termed the hidden curriculum of apartheid, the racial segregation of groups into different school environments was specifically designed to instil notions of separate development and the superiority of white, in particular Afrikaans, culture. There was a political purpose to this ordering of education. In other circumstances, it might be more pedagogical or technological issues that inform the design of educational spaces. For instance, the growth of child-centred pedagogies in the 1960s and 1970s in the United Kingdom resulted in new primary school buildings tending to be designed on more open plan principles, allowing and encouraging a greater fluidity to the interactions for students and teachers. More contemporaneously, the growing use of information ad communication technologies have led to the design of flexible spaces, which individuals use as and when, where a desk and internet access is all that is provided. Thus many influences can inform the particular designs of educational spaces.

However, it is important not to focus solely on the physical and material design of these spaces, as they are not merely a backcloth determining how people interact with and within them and the meaning they take from them. People interact with each other and objects in space and in so doing construct, disrupt, and resist meanings and understandings. They therefore invest certain meanings in their built environments through the forms of interaction in which they engage. Thus, as Karlsson points out, racial segregation was a more complex spatial experience for those who experienced it, compounded by the fact that schools themselves exist within a network of physical, economic, political and cultural relations. Thus, what is learnt is not simply or even what is intended through the formal curriculum, as what is learnt is not confined to what goes on in formal classes, but is part of a wider skein of relationships both within and outside educational institutions. The attempts to spatially order the curriculum is therefore always subject to disruption, because one cannot escape the wider networks within which it is enmeshed.

This struggle for meaning is not restricted to one between teachers and students. It also takes place between different subject teachers, where subject boundaries can be marked by departmental boundaries and the physical allocation of spaces within educational institutions. Carrie Paechter gives an example of this in her case study of the introduction of design and technology into the curriculum of secondary schooling in England, which brought together the previous separate subjects of craft, design and technology (CDT) and home economics. Questions of knowledge and identity are bound up in both the physical and subject spaces that these subjects had previously occupied and were expected to share. Furthermore, given the subjects involved, there were important aspects of the gendered nature of the curriculum spaces, with home economics traditionally a feminine domain and CDT masculine. Thus, even before students enter into educational settings, the curriculum is itself a contested space for those who teach.

Jane McGregor examines the importance of the curriculum subject and its location within a department in more detail in her chapter. Drawing on her research in English secondary schools, she argues that schools are "precarious geographical achievements," which are made and remade through the practices and relations that go on in and around them. McGregor uses a case study of the Science department in one particular school to explore the ways in which people and objects interact in certain spaces within the school to construct an effective and cohesive curriculum. In drawing our attention to the department as nexus she is suggesting that rather than see a curriculum area as sitting within a context of a school, where what goes on "inside" the former is, in part, structured by the "outside" context, we should rather view it in relational terms. McGregor is therefore pointing to the ways in which spatial assumptions, often implicitly, already frame our understanding and practices.

Greg Mannion also puts a relational understanding of learning forward in his chapter. Drawing on his research into children's participation in the changing and use of school grounds in Scottish primary and secondary schools, he argues that such developments disrupt the notion of the outdoors as the space and time for play by contrast with the indoors as the arena for learning. He is pointing to the importance of school grounds as spaces for learning and the ways in which the features of those grounds, who is entering into the process of developing them and how they are used and under what conditions has implications for the learning and identification of children. This is the hidden curriculum of the playground.

What then are the spatial assumptions embedded in conceptions of the curricula? This is the question that Brent Davis and Dennis Sumara address. Drawing on their work with mathematics teachers in Canada, they argue that a particular spatial imaginary informs much curriculum planning. This is derived from the geometry of Euclid and results in notions of curriculum that are ordered in terms of linearity and progression. Learning becomes a set of stages through which one has to progress, something we might associate with a narrow disciplinary curriculum into which one delves deeper and deeper. And, depending on your viewpoint, this deepening might be valued as a good thing or questioned as to the breadth of learning it supports—bearing in mind that breadth and depth are themselves spatial metaphors which can carry values as well as meaning. Davis and Sumara argue that there is a network of Euclidean understandings underpinning much curriculum organization which associated linearity with rightness and normality. They argue for a different geometrical understanding, one based on fractals and the not normal, through which to order the chaos and complexity of actual curriculum practices. They therefore draw upon alternative strands in science and geometry to rethink the spatial patterns of learning. Similarly to Karlsson, they point to that ways in which the spaces of learning are more complex than those of curriculum, but also that curriculum spaces are themselves complex.

This is an issue explored empirically in a study of undergraduate curriculum by Jan Nespor in the United States (USA). He argues that the meanings of instructional events cannot be inferred from curriculum content, course artefacts or the stated intentions of lecturers, but are rather constituted by the spatial networks that conceptualise people, objects and event. In this view of learning, context is central, but rather than being separate from action, it is constituted through actions. Nespor is therefore reinforcing the argument put forward by McGregor that curriculum is a nexus. Here "contexts are not static encircling frames, but are partly presupposed and partly created through the actions of students, professors, and distant participants like textbook writers, admissions officers, and employers." Instead of curriculum being conceived as relatively stable bodies of knowledge, for Nespor it is built through networks of mutually constraining conceptualisations extensive in space and durable over time. This is complex idea, one which points to both the importance of space to understanding curriculum and learning, and to the potential complex understandings of space that exist. Like other contributors to this text, Nespor does not take space to be a static backcloth, a stage upon which human history is acted. He views it is as actively constituted through the people, objects and forms of inter-connections by which these are networked.

Actively constituted, but also experienced. It is the latter that is the focus of the chapter by Peter Gray. Drawing upon his research into student nurse experiences of higher education in Scotland, Gray uses the notions of proximity, mobility and possession to conceptualise their relationship to the spaces in which they learn. In particular, he focuses on the experience of the placement these students take up as part of their courses. The placements may be in medical establishments at some distance from their place of study or their homes, and involve a varying degree of travel. Thus, the nurse students' learning experiences are diversely constituted through their differential patterns of travel and interactions both with other students, but also with existing practitioners in the field. Their proximity to places, their mobility in and between them and their sense of possession of spaces, for example, a place of my own, all impact upon both their learning and their identity. As with many of the contributions to this text therefore, Gray is pointing to the importance of the overall experience of students to their learning and not simply the formal curriculum in formal educational spaces.

Yet these latter remain both important and powerful. As Arthur Wilson and Ron Cervero argue in their chapter on the adult education curriculum in a U.S. University, by placing a programme of professional development within the geographical location of the college, a certain status was conferred on the learning undertaken. Thus, not all learning spaces are valued equally. In a sense it is precisely the struggle to establish which places can be valued as learning spaces of particular types that marks, and is marked by, a formal curriculum. For Wilson and Cervero, the adult education curriculum is best understood as "a social process of negotiating interests in relations of power" and that there is geography of power at play in curriculum planning practices. These are expressed in the very physical location of courses. Curriculum spaces are therefore places that are the outcome of, and give expression to, the distribution and exercise of power.

Nor are these spaces simply constructed by those acting within it. Teachers and students occupy certain types of space, but these spaces have to be managed—for example, they have to be cleaned. In terms of learning therefore there is a complex set of orderings at play. This is illustrated in the chapter by Julia Clarke, which draws upon her research in Further Education colleges in England. Nor are the spaces for learning isolated classrooms. Clarke asked her informants to draw a picture of their most important spaces, which they did, if often reluctantly or with difficulty. These pictures, like Gray's interviews, provide insights into where people learn and the ways in which spaces other than formal classrooms play a central role in such processes. They also point to different ways of researching spatiality in education, although with different approaches come questions about how to interpret and represent the research. Clarke is therefore both outlining her research on spatiality and education, but also reflexively exploring how we can explore such issues. Here what we know depends upon how we know. And, as with curriculum practices more generally, so with research; to learn is to be spatially located and related.

Learning to research has not traditionally been thought of in curriculum terms. However, with the rise of the professional doctorate and the increasing attention being given to research training, curriculum and learning questions within the research process emerge as an important focus for attention. Drawing on her work with research students and supervisors in an Australian university, Pam Green explores some of the spatial dimensions of the increased emphasis on the timely completion of doctoral theses. Put simply, she argues that the requirement to ensure completion creates the conditions for space-time compression of doctoral study. This compression has various dimensions—relational, intellectual and physical—with the result that ironically, even as more doctorates are achieved, the knowledge produced through the research process and the capacities of the researcher to be an improvising performer are diminished. This argument is one that can be made of any outcomes based curriculum of course, but as doctoral work is about making an original contribution to knowledge, the diminishing space-time for the supervision and conduct of research indicates a worrying trend for the type and forms of knowledge which are produced and validated through doctoral study.

A further irony however is that even as there might be diminishing spaces for supervision and research, the distances across which these practices are supported are stretched by the extensive use of information and communications technologies. What types of space are involved when we move into online environments? What differences does it make when the space for learning is a networked and virtual space? Robyn Barnacle draws upon her work with research students in an Australian university to explore this issue. The role and significance of the Internet in curriculum and learning has become an extensive area of debate. Often it is positioned in a slightly dystopian way, insofar as it is seen to distance the relationships between people by bringing them together only virtually. Here there is an assumption that the primary learning community is one of face-to-face relationships. Anything other is seen as representing a diminished form of learning. Others take a more utopian view of the possibilities for globalized learning communities. Barnacle explores these positions and posits a more cautious approach, recognising both the changed relationships supported by the Internet, but also the possibilities for supporting learning communities that are more conducive to the more general condition of globalization.

Central to Barnacle's argument is that information and communications technology should not be seen simply as a technical tool to support learning, but has to be considered for the reframing of learning and experience that it produces. This is also the case with the argument put forward by Michael Peters. Drawing upon Heidegger and Foucault, Peters argues for the need to see that it is bodies that are organized in space and therefore embodied experience is fundamental to learning and identity. This tends to suggest that online learning—disembodied learning—can only be a degraded form of learning, what, taking from Dreyfus, he argues to be the development of competence rather than practical wisdom. This is important as, like Wilson and Cervero, Peters identifies spatial orderings as exercises of power and is keen to support geographies of resistance in critical pedagogy. There is thus a politics of space in critical pedagogic practices that needs to be highlighted when seeking to challenge the exercise of power.

By contrast, one of us—Robin Usher—offers a very different interpretation of the significance of the Internet for learning. His chapter is about the relationship of space and technology, particularly writing as a technology. This wide notion of technology is deployed to consider the contemporary space of the university and in particular the intellectual practices embedded in that space. He argues that this space is both a networked space and a space of "abstraction" and that these are intimately interconnected. New information and communication technologies have been crucial in enabling the growth of the techno sciences and the knowledge economy and in valorizing the mode of performativity associated with these developments. In turn, these have affected the space of the contemporary university to the extent that the form of life of the university and the form of life of the knowledge economy are now mutually interactive. Usher argues that intellectual practice is both an abstracted space and itself constructs a space of abstraction. In this space, social exchange takes the form of "interconnectivity" rather than face-to-face relationality—an interconnectivity enabled by the technological mediation of writing. Technological mediation of any kind creates webs of interconnectivity. The world is reconceptualized as an electronic space, a space of infinite flows. The Internet has made interconnectivity possible on a global scale, connectivity which taken to its fullest extent becomes hyper-connectivity. This hyper-connectivity has implications for the face-to-face and in particular for the diminished importance of relations of presence. With this also come new forms of identity and changes that reinforce or unsettle boundaries and hierarchies. In constructing his particular space of writing, Usher attempts to exemplify his 'message' through a technology not only of writing but also of electronic text production and Internet research.

When we first envisioned this text, we planned to write a conclusion to draw together some of the issues and themes emerging from the various contributions. We have now decided against this, as the reading space opened up in the different chapters is rich and diverse, with connections, interrelations and challenges that do not lead to the type of tidy results so often given in a rounded and definitive conclusion. A text like this has opened a space and the colleagues who have generously given of their time to contribute to it variously populate it. For readers who are interested in connecting with us on any of the issues raised, our contact details are in the back of this text. There remains much to explore and examine in focusing on spatiality, curriculum and learning and many cartographical imaginings to be embraced. The webs of interconnectivity are now in place!

REFERENCES

Comber, C., & Wall, D. (2001). The classroom environment: A framework for learning. In C. Paechter, R. Edwards, R. Harrison & P. Twining (Eds.), *Learning, space, identity*. London: Paul Chapman.

Foucault, M. (1979). *Discipline and punish: The birth of the prison*. Harmondsworth: Peregrine.

Focuault, M. (1986). Of other spaces. *Diacritics, 16*(1), p. 22.

Gregory, D. (1994). *Geographical imaginations*. Oxford: Basil Blackwell.

Harvey, D. (1991). *The condition of postmodernity*. Oxford: Basil Blackwell.

Lawton, D. (1984). Metaphor and the curriculum. In W. Taylor (ed.), *Metaphors of education*. London: Heinemann.

Nespor, J. (1994). *Knowledge in motion.* London: Falmer.
Paulston (2000). *Social cartography: Mapping ways of seeing social and educational change.* New York: Garland Publishing.
Peters, M. (1996). *Poststructuralism, politics and education.* London: Bergin and Garvey.
Tomlinson, J. (2000). *Globalization and culture.* Cambridge: Polity Press.
Wright Mills, C. (1983). *Sociological imagination.* Harmondsworth: Penguin.

CHAPTER 2

REMEMBERING APARTHEID

Political Discourse in School Space

Jenni Karlsson

The apart-ness of South Africa's *apartheid* system was a literal spatial separation of people on the basis of race. The legal framework for this system took shape following the 1948 whites-only electoral victory of the National Party, and only began to be dismantled in the early 1990s when the National Party conceded the need for non-racial democracy. Legislation of the apartheid era (1948–1994) encompassed the racial identification of every person, racialized residential zoning, and restrictions on movement and employment. State schools were an important part of institutionalizing racial segregation and for 'fixing' every child's racial identification. Each school was geographically located to serve only the racial category for which the area was zoned. Fourteen education departments administered and provided personnel, resources and curriculum for the spatially segregated schools.

The spatiality of apartheid has been much discussed. For example, Davies (1981) theorized the apartheid city and Robinson (1996) formulated an argument about the "location strategy" of the Apartheid State's dormitory townships. These were planned and developed on the periphery of the apartheid city for the sole use of indigenous Africans, initially as a temporary base for workers who would return to their rural homes. Fur-

Space, Curriculum, and Learning, pages 13–27

ther, the exhibition *blank Architecture, apartheid and after* (Judin & Vladislavic, 1998) brought together the understanding that many South African architects and urban planners now share of apartheid-era political use and abuse of built environments, including educational institutions. Education analysts, however, have not addressed apartheid spatiality in schools. For example, during the apartheid era critical scholars such as Kallaway (1984), Nkomo (1990), Hyslop (1990) and Unterhalter et al. (1991) focused on the systemic inequalities and institutional differentials evident in racial comparisons of per capita funding, enrolment and curriculum, as well as modes of popular resistance.

It is only in the post-apartheid era that some education analysts have turned their critical gaze to the discursive organization of school space. For example, Jacklin (2001) is engaged in a study of pedagogic and management practices in school contexts, and I (Karlsson, 2002) have examined the material conditions at diverse schools and how these continue to reflect and perpetuate educational inequalities. This shift is largely because the determinant institutional imperative in post-apartheid South Africa is the pursuit of social equality and the need to "level the playing fields" of schools. On the whole the physical infrastructure at schools has remained relatively unchanged since 1994 and stands as a reference and memorial to the apartheid past. This is despite government policy that funnels State funds towards previously disadvantaged schools in rural areas and towns. Social class therefore appears to have replaced race as the key social differential in South African schooling (Pampallis, 2000; Motala & Porteus, 2001). Other analysts and commentators assert that apartheid's spatiality will be felt for many years to come (Davies, 1990; Frescura, 2001; Matshikiza, 2002) and this appears to be confirmed by the slow progress noted in the 2000 *School Register of Needs Survey* (Motala, 2002). Nevertheless, it is by understanding how the political discourse of apartheid worked in school space that South Africans could ameliorate what's left of it. For this, a starting point is to explore how individuals subjectively experienced apartheid discourse and how school spatiality was appropriated for that project. It is the exploratory work of digging in, sifting through and looking at memories of school that is central to this chapter.

METHODOLOGICAL AND THEORETICAL CONCERNS

Like an archaeologist searching for fragments and shards that might help reconstruct a meaning of the past and how things were organized, I have worked with memory accounts from eight people who attended schools during the apartheid era. My intention was not to understand their personal life histories and self-identification, but rather to draw out and piece together, which places at school were meaningful as a shared history about learning apartheid. The data was drawn from a larger study about apartheid and post-apartheid discourses in school space. Notwithstanding the subjective embroidering and selectivity in reminiscences, the personal nuances and associations, memory accounts can illuminate and form the basis of our understanding of an historical period (Engel, 1999). Memory work has value in the post-apartheid era. At the public forum of South Africa's Truth and Reconciliation hearings, the production of memory was promoted and criticized as a lever for social reconciliation and collective healing (Nuttall & Coetzee, 1998). The flood of autobiographies published in the 1990s also signal the important role remembering has had for individual South Africans to tell their stories and make sense of their apartheid experiences.

My theoretical starting point, that school space provides a field of operation for a political discourse such as apartheid, rests heavily on the work of Lefebvre (1991). He asserted that space is socially constructed and used politically to produce social relations. His point seems so obvious that it is often overlooked. It is that the built environment of a school site has neither developed naturally as an act of God, nor is it value-free and socially neutral. Indeed, public schools in South Africa, past and present, are designed by State architects and built according to norms and standards determined by the authorities, to effect schooling as given in social policy. Apartheid-era educational sites were concrete manifestations of the values and policies of apartheid that "sought to prevent the intermingling that is a mark of the modern" (Herwitz, 1998: 417). Citing examples of university campuses Herwitz (1998: 415–416) has argued that "the architecture of the apartheid regime ... concern[ed] itself with who comprised the South African nation and how it would live, with the rules of identity and citizenship ... buildings of extraordinary size and grandiosity, the built expressions of an unbridled power". Simultaneously the State neglected racial communities falling outside the apartheid notion of full citizenship, throwing those black, "colored" and Indian communities on their own resources for schooling. Such school sites, mostly in rural areas, inevitably reflected local practices and poverty conditions, and took no account of norms and standards regulated by some distant State bureaucracy.

That apartheid was about race is common knowledge. However, respected South African scholars such as Wolpe (1988), Norval (1996) and Posel (1991) have argued that apartheid was about more than this. There is increasing consensus that apartheid operated through a nexus of race, social class, space, culture and patriarchy to stratify South Africans within a hierarchical system for privilege or subjugated disadvantage. How did those attending school learn the stratification of apartheid? How do they express their memories of that social order? What places were significant for their learning to accept their position within the stratified hierarchy as the natural social order? And how did individuals use their school space to resist apartheid?

The eight adults whose memory accounts I draw on in this chapter comprised a heterogeneous group. In terms of racial identification, two were black (indigenous Africans), three were Indian, their forebears being from the Indian sub-continent, and three were white, of various European settler origins. The informants were purposively sampled to exclude "coloured" and Afrikaans-speaking white informants since these communities are not numerically significant in the city of Durban, the locus of the study. Unlike other areas of South Africa, Durban's white community is predominantly English speaking, a referent to Durban's history within the former British colony of Natal. There were five women among the group. Although they all spoke English and interviews were conducted in English, it was the first language for the Indian and white informants only, with isiZulu being the first language of the black informants. In accordance with the policy of apartheid, all informants had attended racially segregated schools administered by education departments for schools only serving particular racial groups. One black and two white informants attended schools with boarding facilities, and two white informants had attended single-sex schools. The Indian informants attended schools relatively near their homes. Notwithstanding these schooling differences, the group shared some characteristics. Not only had they all taught at some time at schools in the city of Durban, they were in the 30–50 year age group and had lived through the period that included forced removals and urban relocations of the 1960s and 1970s and the Soweto uprising in 1976.

LEARNING YOUR PLACE

One of the striking features about how apartheid was learnt is that racial segregation really was effective in schools and kept learners apart. The system's spatial and administrative segregation of race-based schooling hermetically sealed learners from encountering their peers of other races, so that there was a vacuum of knowledge and understanding about the Other. One white woman, Lucy, said that contact was only possible for her once she reached University. This was confirmed in Jay's account, in which she only recalled encountering people of other races as domestic workers, clerks and laboratory technicians at the whites-only schools she attended. Similarly, in her entire schooling Laila, an Indian woman, only met white learners when she participated in an inter-school debating competition. And she made no mention of contact with any black learners in her memory account. Another Indian informant and active sportsman, Mike, stated that he had never had any contact with schools and learners outside of his own neighbourhood when he was growing up, and that league events were restricted to Indian schools. Although Lindiwe, a black woman, attended a former mission secondary school that employed many white teachers, mixing with pupils of other racial groups was beyond her wildest dreams: "we wouldn't go to any white school. It was only white teachers that taught us who were white, but otherwise with white schools altogether, no Indian schools, no coloured schools, no mix. That was a dream [laughs] which came true when we got the freedom [from apartheid]".

Given this almost total insulation of schools and administrative authorities within the apartheid education system, that prevented learners of different racial groups from learning together and even engaging in extracurricular activities and competing on the sports field, it is no wonder that informants' young notions of social hierarchy are not mostly about race, but instead foreground cultural and social class differences. Vishnu articulated poignantly his perceptions of social difference when he encountered other Indian learners at a public library:

"A lot of those [Indian] pupils [...were] from very affluent homes in town and in a sense their lifestyle was a lot different from those of us that came like maybe from an outlying area.... Those kids and the schools that they come from are very different from where we are and what we hold near and dear to us. So one could feel a kind of gap between us....You see, a lot of people that come from those schools belong to the Muslim community, the Gujurati community and so on, and in a sense there are differences in lifestyle and ... you find that you come from these institutions and areas where you have working class parents and so on, and these are businessmen, children of businessmen, and much affluent people. So you could see [the gap] in, I think, different ways, in the dressing, just in the sense of superiority and arrogance that one could detect.... They have had the opportunity for greater achievement in different fields. They even played different kinds of sport that we never even had the opportunity of.... Straight away you can see and maybe as you say, the kind of confidence, possibly because that is also their own environment, that library is something that they visit on a daily basis so they have a greater affinity and greater knowledge about things that go on around them. But for us [the library] was like coming to a strange environment."

Vishnu's account reveals a complex calibration of privilege within Durban's Indian community in which he constructed middle class and Muslim Indian learners as the Other and against whom he felt inferior and disadvantaged. His remembrances also gives insight into the way apartheid forced the erasure of heterogeneous groups by collapsing their official identification under one rubric (Ebr.-Vally, 2001). Howarth (2000a, 2000b) and Norval (1996) refer to this logic of equivalence as a technique used systemically in the hands of the apartheid authorities to assert the white supremacist identity and hegemonic power over Others.

In one of Lucy's memories she made similar observations to Vishnu, but for her the differences were about linguistic and cultural cleavages within the white community. The time she recalled was when she attended schools in Pretoria, the executive capital of the South African government, once the symbolic heartland of white supremacy. Her learning about apartheid social order sprang from a linguistic dichotomy of English/Afrikaans white learners that created barriers even in institutions: "I remember there was a lot of animosity between Afrikaans schools and the English schools. We always- for one thing we had no interaction.... I don't often remember speaking to people from Meisies Hoër [=Girls High] and Seuns Hoër [=Boys High]. They were like other creatures. They just didn't interact. We were quite resentful of them because we saw them as [...] Nationalist kids." By attending an English language school not only was Lucy isolated from her cohort of other races, the cultural divisions among schools for white learners alienated her from her Afrikaans-speaking peers. For her, white Afrikaners were the Other, "like other creatures," and worthy of resentment because of the subjugating power they represented by dominating the ruling political party of the apartheid system.

Like Lucy, the high school that Angela attended in South Africa was for English-speaking white learners only. She conceived of her society in terms of divisions among whites only. But her perception of hierarchy refers to spatial practices that fractured along social class rather than linguistic lines:

> "[Our school was] a dumping ground for problem children from Durban.... Hostel groups stick together and the day scholars stick together in terms of area where you live, particularly in a big high school like that you tend to group yourselves. I would never have made friends with one of those [hostel] pupils. They were not just- socially they just were very different to me.... They were already going to discotheques and using makeup and they were far more advanced socially, I think, than I was.... I think they were more experienced in terms of sexual relationships and that type of thing...."

By defining boarders at the school hostel as "problem children" who behaved in ways prohibited within her family's middle class culture, Angela constructed the school boarder as the Other, and she used the school hostel as the organising principle to maintain her concept of social hierarchy at school.

Bheki and Lindiwe had similar memories of school experiences that brought to the fore how they learnt about social class distinctions. As was the case with other informants in their racially insulated school world, for Bheki and Lindiwe their memories concerned the black community.

Their distinctions were not only economically indexed. They stratified learners along a rural/urban spatial divide, thus exposing a complexity within the black community ignored by apartheid's racial classification system. When he went on excursions to rural schools Bheki, who attended a township school, identified himself as having a strategic advantage over his rural cohort: "we used to go to rural areas where I will say our uniform was perfect compared to the rural areas. Mind you, the rural areas there's no running water, you see. And then in some excursions we used to go to schools, which were situated in urban areas. They used to wear no different [uniform] comparing this with mine." In this fragment we see that through his school excursions Bheki had begun to learn that apartheid's hierarchy included an urban advantage. But his spatial insulation in the city that closed him off from seeing the superior privileges of white learners prevented him from going further in his critical understanding of apartheid spatiality. Thus, he stopped short of expressing a negative perspective on the State's miserly township investment, which was merely a sop to ensure a minimal level of docility and stability among the labour force (Robinson, 1996).

Lindiwe was from a middle class black family, and both her parents were education professionals. In their endeavour to give her a good education, they chose to send her to boarding schools. Her memories of boarding school and holidays at home in the township mapped further complexities within the urban/rural divide to include an ambiguously positioned black middle class that, like their white counterparts (Morrell, 2001), used rural boarding schools to school their offspring. Sometimes Lindiwe associated middle class with being urban, but when in the township she distanced her middle class practices from loitering urban youths:

> "...That boarding school was in a rural place and people from rural areas don't even care whether the child is at school or not. Eh, absenteeism is kind of day [scholar] thing ... Kids can drop out any time, you know. The community they have doesn't have the sense of doing things, unlike at A. School. At A. there they were known for loving education.... During holidays I used to go back to the township. The kids I used to go to [day] school with, you know, they kind of distant themselves from you, I think maybe even to themselves they say 'oh they are now better than us' and the child from the boarding school is used to staying home, watching TV. You know you don't just go and loiter around streets like township life.... When you are from a boarding school you do get resent because most of the time your life has been structured and it has got a programme."

Here Lindiwe structured a social order, but it is not about racial hierarchy. Instead her reminiscence brought to the fore social class and space, thereby undermining the apartheid-era reductionist myth of a monolithic black population. Lindiwe did this by juxtaposing conscientious middle class boarders from urban centres against dropout day scholars from poor rural areas. The exposition is textured in Lindiwe's spatial contrast of the unfocused, spontaneous life on the township street and the controlled, technology-infused schedule that confined her middle class home life.

On the whole, the memory accounts show that while the insulation around segregated schools masked the most significant apartheid division of race, nevertheless the structuring of hierarchy in the social relations of these learners had been learnt and was operational albeit along axes of class and culture. Arguably these findings from a few South Africans might mirror the quotidian of learners anywhere in the world. But the significance here is twofold. First, these memories bring to the surface the way everyday school experiences are implicated in and support the internalized structuration of social hierarchies. Schooling failed to provide these learners with an alternative compass point for understanding and organizing society other than a hierarchical nexus of economic, cultural and spatial privileges. In the second place, these memories point to a learnt social order that provided the branch for grafting on a racist hierarchy as the natural order.

There were rare occasions in the memory accounts that provide windows specifically on how South Africa's radicalized social order was being learnt at schools. One explanation, the spatial segregation, has already been discussed. A memory fragment from Jay, a white learner, provides another explanation. Censorship and threatened censorship were widely used by the authorities to prevent the populace from developing an understanding of the full extent of oppression and atrocities within the system (Merrett, 1994). Self-censorship, an indicator of the State's success with instruments of intimidation, emerged in Jay's anecdote of a senior year in secondary school: "… I put up … photographs from magazines, Time-Life magazine. I had put up a picture of a mixed couple, the photograph was in black and white, and I was told to take it down." The ruling from her teacher, a State employee whose job might have been threatened had she not removed the photographs, taught Jay that even imagining racial integration was disallowed. Censoring such images in white schools ensured that white learners never constructed racial integration as "normal" so that they continued to internalize the peculiarities of the white South African supremacist myth of *swart gevaar* (black danger).

Vishnu's account reveals that, despite segregation, by grade 8 he conceptualized social order as a white/black dichotomy but that within his notion of "black people" there was a further calibration of racial privilege: "…white people being the rulers and the black people being everybody else, that is not enjoying the kind of privilege and the society that one would have … and having this very deep sense of the African society and

African people …the worst off." The racialized order that privileged "white," discriminated against all other races, while assigning relative advantage to "Indian" and "coloured" over the category of "black" African (Ebr.-Vally, 2001). The 1983 Constitution that established a tri-cameral parliament for whites, Indians and coloureds, but excluded blacks, formalized this order systemically. Cultural and spatial relations decided the logic of this development of the apartheid order (Norval, 1996; Tikly, 1993). Urban centres were constructed as "white" and controlled by white authorities, while Black people were assumed to have permanence in a rural homeland and only temporary workers lived in cities and towns. The 1983 Constitution formalized the pre-existing manifestations of gradation evident in Vishnu's account. Thus, he interpreted apartheid as a complex in which the Indian was below whites but not "worst off," a position he perceived was reserved for indigenous Africans.

From all these memory fragments we see that learning apartheid was not a matter of objectifying and being objectified simply by the indicator of phenotype. The apartheid order was internalized through direct and censored experiences that allowed a complex and nuanced differentiation of interlinked race, class and spatial relations.

PLACES FOR LEARNING APARTHEID

The memory extracts discussed above are remarkable in that school lessons and classrooms did not emerge as necessarily essential for learning apartheid. This is important because, in much of the literature about apartheid education, the curriculum of fundamental pedagogics and Christian National Education are criticized for enslaving the minds of South Africans. However, in the memories of the eight adults there are anecdotes about other places *en route* to and from school and beyond their own schools that shaped thinking about the apartheid social order and positioning oneself within the racial, class and cultural hierarchy.

In the public sphere of the street and on public transport travelling home at the end of a school term, Lindiwe found racial difference brought to the fore. That was where she learnt her most acute lessons of being black and that race was the primary measure of a person in apartheid South Africa:

> This racial thing, the hatred about this racism would only be felt when we supposed to get into public places where you were not allowed to buy from and get into this restaurant. You are supposed to peep through the window to get something you need. You were not supposed to get into first class of a train. You were supposed to use the third class. Those are the kind of things to make you feel that, 'ay, I'm black now'.

To be black was a negating experience. Although Lindiwe had not changed phenotype from when she was in class with her white teachers, she found that the movement from segregated school space into the public sphere of intermingling apartheid society diminished and restricted her social status and stature. Her expression "I'm black now" reveals the shifting consciousness of socially constructed identity that moved rapidly from"learner" identification to an acute moment that reduced her to phenotype identification only. This experience highlights the importance of the public sphere as a place for learning apartheid. It was on buses and trains and on streets travelling to and from school that limited racial contact may have occurred and acted as a fracturing in the insulation of the segregated school system.

The significant contact zone (Pratt, 1991) for Laila was beyond her own school, when she visited a privileged independent school in an affluent white residential area in a wooded part of Durban where there were large residential plots, smallholdings and places specializing in equestrian sports. The occasion for Laila's visit to that school was a debating competition:

> "I was on the school debating team, I was eventually chosen to represent the school in some circuit where we came up against other schools. It was hosted at [an independent school]. The experience of going there with the teachers and going into the [school] and looking at the buildings and the school hall. It was amazing. Just to see the contrast.... I was aware that I came from a disadvantaged institution ... I became very despondent about having so little."

Overawed at her "discovery" of race-based discrimination in the physical space of school, Laila deemed her school, with facilities, which she had earlier described in superlative terms, to be "disadvantaged." Through experiencing school spaces that extended beyond her Indian working and middle class schools to include a superior school space designed for a white elite, Laila's earlier inflections of class as the principle of social order became eclipsed by race. While in earlier reminiscences Laila had referred to her endeavours at school to ensure her upward mobility along the axis of class, her devastation in this extract suggests that through this visit she comprehended the finality and hopelessness of any classification less than "white" on the apartheid axis of race. It may have been some years before Laila learnt that lesson about apartheid had she not visited that school for the debating competition.

There were some memories, also outside the classroom, specific to white informants that showed rituals and mass parades were used to promote the nationalism and triumphalism of apartheid among white learners. It was in the school assembly hall that Lucy recalled being subjected regularly to rituals that blurred boundaries between religion and national symbols of flag and anthem. It was through such repetition at school that white learners absorbed nationalistic loyalty to the Apartheid State.

Participation as a school representative at provincial celebrations of the apartheid republic's 5th anniversary might be an unlikely site for learning about apartheid, but it was one that Jay remembered:

> "There was going to be a huge big celebration by the province in Pietermaritzburg in a massive big stadium ... every primary school, every high school, was sending representatives ... I was the 'lucky' one who had to go and sit through this dreadful experience in Pietermaritzburg, sitting shivering in a cold stadium and having to walk around the field holding the school banner with this boy that I didn't really like... and even I can remember the song that they sang in the evening and there were fireworks..."

Despite her own particular account of hollow white triumphalism, Jay's reminiscence gives us the opportunity to see how white learners were gathered together occasionally for spectacular mass events in stadiums. Although such events were infrequent, the stadium provided white school learners with the collective imprint of "buildings of extraordinary size and grandiosity, the built expressions of an unbridled power" (Herwitz, 1998, pp. 415–416) that the apartheid regime achieved through the architecture at some higher education institutions.

PLACES FOR RESISTING APARTHEID

While the extracts above about learning apartheid convey the impression that young learners passively assimilated and accepted apartheid as a natural social order, evidence from the memory accounts show that learners were not necessarily and/or entirely passive. Much has been written about mass-based struggles against apartheid (for example, see Cross, 1992; Kallaway, 2002; Unterhalter et al., 1991), but these have largely concentrated on resistance from black learners. Very little is known about the comparatively inconsequential, undramatic and highly personalized resistance among learners who did not take to the streets. Yet in the eight memory accounts under discussion here, three forms of resistance surface: defiance against a school assembly nationalism ritual; raising current political issues in a written assignment; and in a display on a classroom pin board.

Though these forms were subjective, individualized and unremarkable in the significance of their overall effect on the apartheid system, they were sufficiently meaningful for the individual to have been remembered at all. It is in the intimacy and modesty of the personal resistance through which we see the pervasive intrusiveness that apartheid had on the everyday life of all South African learners. Further, these small signs of resistance attest that Fascism, violent suppression and racism were, indeed, in the consciousness of apartheid-era learners.

Lucy's defiance manifested in modest, almost imperceptible, body movements. In the school hall at assembly she was required to stand for the former flag of apartheid South Africa and sing the anthem in Afrikaans. She clearly recalled "hating having to stand for the school—for the South African flag—and tried to sort of drag [my] feet and stand and not sing *Die Stem* (the former national anthem of South Africa)." The coerced nationalistic patriotism of a white school assembly led to her defiant attitude and posture. She had become adept at self-censoring her defiance and reduced her movements so that only she was wise to her meaning, thereby avoiding detection and punitive measures. Thus, even in the ritualized space of the school hall, the cover and anonymity in the mass of her cohort provided conditions for a modest spatial practice of resistance.

Jay also made provocative attempts, as discussed above, to decorate the notice board in her classroom with images of forbidden relationships across the colour line. Although the outcome of her challenge was unsuccessful and she was required to remove the magazine pictures, like Lucy's account of resistance, it demonstrated the political nature of white schooling where discussion of race was discouraged and learners learnt a manner of acting blind to the problematic of apartheid society by accepting the racist hierarchy as the natural order.

In contrast to the control over debate and gaze at white schools, at his Indian school Vishnu found more latitude to express thoughts about the atrocities of apartheid.

> "When I was in [Grade 11] it was the 1976 riots, I think for the first time I was able to articulate a lot of what was happening 'cos there was a piece of work that I did write for a composition within school. So I was able to put it down in writing although the concepts may not have been very clear even at that stage. But one could see that there were those differences in society."

He used his composition notebook as a place to engage with the issue of racial difference in a South African context. As assignments, essay compositions require individual expression and perspective. This gave Vishnu a private yet public space where he felt sufficiently secure and permitted to venture into a discussion of race and South African society. His silence about the teacher's reaction implies that he was not censured for using his schoolwork to register his voice of protest.

While the apartheid system sought to regulate all aspects of public and private life, from the experience of Lucy and Vishnu it appears that at schools more private modes of protest and defiance were tolerated or went undetected. However, the public sphere of the class notice board that Jay had sought to appropriate was a space that she found to be controlled and subjected to censorship. She found no latitude in the teacher at her white school to allow a representation of an alternative to the apartheid way of life. Such a public display would have been too dangerous, too infectious.

CONCLUSION

To sum up, the subjective and reflexive memory accounts of the eight adults in this study who attended school during the apartheid era have shown that learning apartheid was for all South Africans an intimate personal journey. Their school years were significantly formative for identifying that society was structured hierarchically and for concluding the social consensus of their position on that ladder of racial, social class and spatial privilege or disadvantage. The memory accounts also reveal that this was a lesson rarely learnt through a crude doctrinaire approach in formal classroom lessons, although I am not suggesting that this might not have occurred at some schools. Assemblies, inter-school events and parades, as well as commuting daily to and from school were among the critical, sometimes epiphany moments remembered for their specific discursive content about apartheid's racial segregation, discriminatory oppression and stratified social structure. Also in these accounts we find that outside the momentous political upheavals of South Africa's social history, in their lived experience of school space, young people across the racial spectrum found their own openings and outlets for a personal form of protest.

Although South Africa and apartheid discourse were central to the discussion in this chapter, the issues are of wider international significance, especially in countries where democracy is still weak or even absent and where schools are used instrumentally to advance the interests of a minority. This is because outside the formal curriculum, there are many things that children learn about society and themselves at school. Spatiality is an important component of that non-formal learning.

Underpinning the enquiry of this chapter was the theoretical notion, simply put, that State policies influence and activate the spatiality of public schools in keeping with the social order and to maintain the dominant political discourse. Corresponding to that notion was the assumption that space, even school space, is a field filled with discourses that inform and shape social relations. The lens of apartheid and South African public schools was used to examine aspects of that dialectical process involving space and learning the hidden curriculum. How school space effected this has not been well understood, but the memories of apartheid-era school life showed that classrooms were not a significant space. Instead, it was in spaces such as the street between home and school, in places like school assembly halls and contact zones like excursion and competition visits to schools where the Other was encountered that were meaningful. Further, memory accounts showed that learning apartheid was more than learning the racial ordering of South African society; it was also about social class and the urban/rural divide.

Although we may now understand more about how apartheid discourse was spatialized and internalized by children in and through space, Lefebvre's contention that old spaces are not appropriate for new spatial practices turns our attention to post-apartheid discourse. For this we will watch how South Africa's post-apartheid decision-makers reshape school space and spatial practices for a transformation discourse of democracy and equity that values all learners equally, regardless of their race, social class and geographic location.

REFERENCES

Cross, M. (1992). *Resistance and transformation: Education culture and reconstruction in South Africa.* Johannesburg: Skotaville Publishers.

Davies, R. (1990). Post-apartheid scenarios for the Southern African region. *Transformation,* 12–39.

Davies, R. J. (1981). The spatial formation of the South African city. *Geojournal Supplementary Issue,* 59–72.

Ebr.-Vally, R. (2001). *Kala Pani: Caste and colour in South Africa.* Cape Town: Kwela Books and South Africa History Online.

Engel, S. (1999). *Context is everything: The nature of memory.* New York: W. H. Freeman and Co.

Frescura, F. (2001). The spatial geography of urban apartheid. In A. Zegeye (Ed.), *Social identities in the new South Africa: After apartheid.* Cape Town: Kwela Books and SA History Online.

Howarth, D. (2000a). The difficult emergence of a democratic imaginary: Black consciousness and non-racial democracy in South Africa. In D. Howarth, A. Norval & Y. Stavrakakis (Eds.), *Discourse theory and political analysis: Identities, hegemonies and social change.* Manchester: Manchester University Press.

Howarth, D. (2000b). *Discourse.* Buckingham: Open University Press.

Hyslop, J. (1990). Schools, unemployment and youth: Origins and significance of student and youth movements, 1976–1987. In B. Nasson & J. Samuel (Eds.), *Education: From poverty to liberty. Report from the second Carnegie inquiry into poverty and development in Southern Africa.* Cape Town: David Philip.

Jacklin, H. (2001). Social space, context and abstraction. Paper presented at the *Social Theory* conference, University of Sussex, Brighton.

Judin, H., & Vladislavic, I. (1998). *Blank _ architecture, apartheid and after.* Cape Town: David Philip.

Kallaway, P. (1984). *Apartheid and education: The education of black South Africans.* Johannesburg: Ravan.

Kallaway, P. (Ed.), (2002). *The history of education under apartheid, 1948–1994: The doors of culture and learning shall be opened.* New York: Peter Lang Publishing

Karlsson, J. (2002). Looking at apartheid school spaces. In P. Kallaway (Ed.), *The history of eduction under apartheid, 1948–1994: The doors of culture and learning shall be opened.* New York: Peter Lang Publishing.

Lefebvre, H. (1991). *The production of space.* Oxford: Blackwell.

Matshikiza, J. (2002). An epitaph of smoke and mirrors. *Mail & Guardian* (June), 21–27.

Merrett, C. (1994). *A culture of censorship: Secrecy and intellectual repression in South Africa.* Cape Town: David Philip.

Morrell, R. (2001). *From boys to gentlemen: Settler masculinity in colonial Natal 1880–1920.* Pretoria: University of South Africa.

Motala, S. (2002). School register of needs survey 2000. *Quarterly Review of Education and Training in South Africa* 9(1),10–13.

Motala, S., & Porteus, K. (2001). School reform in South Africa: Surviving or subverting the system. Paper presented to the Comparative International Education Society Conference, Washington, USA, March.

Nkomo, M. (1990). *Pedagogy of domination: Toward a democratic education in South Africa.* Trenton, NJ: Africa World Press.

Norval, A. (1996). *Deconstructing apartheid discourse.* London: Verso.

Nuttall, S., & Coetzee, C. (Eds.) 1998. *Negotiating the past: The making of memory in South Africa.* Cape Town: OUP.

Pampallis, J. (2000). Education reform and school choice in South Africa. Paper presented at the *School Choice and Educational Change Conference,* Michigan State University, East Lansing, MI, March.

Posel, D. (1991). *The making of apartheid 1948–1961: Conflict and compromise.* Oxford: Clarendon Press.

Pratt, M. L. (1991). Arts of the contact zone. *Profession 91,* 33–40.

Robinson, J. (1996). *The power of apartheid: State, power and space in South African cities.* Oxford: Butterworth-Heinemann.

Tikly, L. (1993). Levelling the playing fields whilst moving the goal posts: Changing discourses on 'race' in apartheid education policy. Paper presented to Kenton at Olwandle Conference, Johannesburg, November.

Unterhalter, E. et. al. (1991). *Apartheid education and popular struggles.* Johannesburg: Ravan.

Wolpe, H. (1988). *Race, class, and the apartheid state.* Paris: UNESCO.

CHAPTER 3

TERRITORIALITY, INTER-DISCIPLINARITY, AND SCHOOL SPACE

Carrie Paechter

Space and time are inextricably linked in schools. The timetable, despite its name, is simultaneously a spatial representation of blocks of time, and a temporal representation of the use of space. As a paper document, in which time is chopped up into boxes of determinate length and parceled out among teachers and between subjects, the timetable represents time using two-dimensional space. In constructing a timetable, pieces of paper or pins or other symbols are physically moved around until they fit together in the completed jigsaw that is the school week; these spatially organized symbols represent temporal blocks in which particular groups of students have to be taught particular subjects.

However the timetable also and simultaneously represents space through its organization of time. It is concerned not just with when things are taught, but where. In a finished timetable, this becomes obscured, but in the timetabling process it is all too obvious; you cannot put a science lesson into a particular time-slot unless there is also a laboratory available in that slot. Most timetables, because of this bringing to the fore of time over space, also treat time-slots as being both discrete and capable of instantaneous change from one to the next: it is lesson one, a bell rings, and immediately it is lesson two. That this has no bearing on any understanding of

Space, Curriculum, and Learning, pages 29–43

"reality" becomes very apparent in multi-site schools where the inability of teachers and students to transmit themselves instantaneously from one space to another becomes starkly apparent; in these cases temporal issues often have to give way to spatial priorities.

While the overt focus for teachers is on the time aspect of the timetable—which one is teaching when—the spatial underlies this in significant ways, which are often related to power/knowledge claims. Powerful departments often have well-resourced suites of rooms, allowing their members to remain in one place for all their teaching and supporting communication and collaboration between their members (McGregor, 2002). Individuals and groups that are less powerful may be teaching all over the school, moving books and equipment from room to room.

Power relations between subject departments can be tracked to some extent in terms of changes in school space use. For example, in my own experience as a mathematics teacher, the changing fortunes of one department in which I worked were directly related to the department's status in the school compared to the other two of the "big three," science and English. The departure of a strong head of mathematics, and his replacement with a significantly weaker head of department, coupled with the arrival of strong heads of English and science, resulted in the mathematics department losing several of their teaching rooms and being increasingly spread out over a large campus. At the same time, a new science block was built and equipped and the former laboratories converted to a specialist suite of rooms for English teaching. Power relations within subject departments also have a significant impact on who gets to use which spaces. In a different school in which I worked, there were only eight places at the staff room (faculty lounge) "mathematics table" for ten members of the department; two individuals had their marginality compounded by being denied access to this departmental space. In many schools in the UK there are not enough classrooms for each teacher to have their own specialist and personal teaching base. This can mean that senior departmental members spend all day teaching in one room, while those more junior wander from room to room with piles of books and equipment.

In consequence, teachers individually and collectively feel very strong ownership of school spaces, from specific chairs in the staff room to particular classrooms or areas of the school. When they come to work across disciplines, the spatial boundaries between these can therefore be as important as subject-based barriers. Interdisciplinary work brings to the fore both power/knowledge relations and spatial issues, as teachers working together have to share both their knowledge and their teaching spaces. In this chapter I am going to demonstrate how disputes about space use involve power/knowledge relations and are used as part of teachers' resistance to interdisciplinary work, and that, because it is bound up both with space and with power/knowledge relations, gender is an important factor in this.

DATA SOURCES

The data on which this chapter is based comes from two interlinked studies of the formation of design and technology (d&t) departments in England in the early 1990s.[1] The 1998 Education Act established this new subject, which, in secondary schools, was to be formed from a group of craft and vocational subjects: home economics (food and textiles) (HE), craft, design and technology (woodwork, metalwork and plastics) (CDT), and, in some schools, business studies and art and design (DES/Welsh Office, 1990). The focus of both studies was the negotiation of the new subject area between teachers of strongly gendered, marginal subjects, whose previous contact had been minimal. These teachers were expected to come together to forge a new, unified d&t department, usually with a single head of department from one of the contributing subjects, and with at least some interdisciplinary teaching, based on a supposedly common "design cycle" (National Curriculum Council, 1990). The focus of the studies was not on the spatial, but spatiality turned out to be a strong theme. The ways in which space was used to sustain and contest the gendered power/knowledge relations between teachers and between their departments of origin was a key factor in sustaining or undermining the good relationships between teachers that are essential for the success of interdisciplinary work (Paechter, 1995b).

SPATIAL ISSUES IN INTER-DISCIPLINARY D&T

For teachers of CDT and HE, the main contributory subjects to d&t, space was already very important. As craft/vocational subject teachers, they used specialist spaces for their work, with woodwork and metalwork shops, textiles rooms equipped with sewing machines and large tables for cutting out, and kitchens for food preparation. Many HE departments also contained a model flat, used in the 1960s and 1970s to train "less able," usually working-class, girls in housekeeping; this had in most cases been converted into departmental offices, which sometimes meant that HE had the best office space in the school. At Bursley School, however, the flat's dining room was still used for the entertainment of visitors (for whom the department had until recently cooked elaborate meals). CDT departments tended to have offices that were also stores for specialist tools, wood, sheets of metal and more dangerous items of equipment; a technician shared between all staff and rooms in the department generally organized these. Most of the departments studied had all their CDT rooms together; at Longshaw Girls', however, the subject had only recently been introduced and was taught in a row of "temporary" classrooms.

Because of the importance of specialist spaces to the teaching of their subjects of origin, for teachers contributing to the new subject of d&t, ownership of space was symbolically related to the nature of their subjects. Without these specialist spaces it was more or less impossible for the subjects to be taught. The spaces were also perceived as potentially dangerous, because of the equipment used in them. This led teachers to impose additional forms of discipline on students within them, to try to set limits on class sizes and in some cases to attempt to exclude certain categories of student. At Turnhill School, for example, a long battle was waged (and lost) over the inclusion of partially hearing students into CDT workshops. Staff argued that they could only teach these students if they were given extra support, on the grounds that the students could not hear urgently shouted warnings. Thus, even before the introduction of d&t, the spaces used by the contributing subjects were both symbolically owned by these subjects and strongly defended by them.

Against this background, the introduction of d&t as an interdisciplinary subject meant that some schools or newly appointed heads of department desired or required teachers to move between spaces for some of their teaching. This might be for a number of reasons. First, d&t departments were now regarded as whole entities from a staffing point of view; if, for example, there were enough teachers in the department but insufficient textiles specialists to cover the curriculum, teachers of other d&t specialists could be expected to fill the gap. This happened at Knype, where a home economist moving to another school was replaced by a specialist in resistant materials; this meant that some of the textiles teaching had to be done by former CDT staff (who were trained on the job by members of the former HE department). In other schools, attempts were made to demonstrate, particularly to students, that this was now a unified subject, and some felt that the best way to do this was for teachers to work across all the contributing areas. In some cases this was welcomed, while in others it was strongly resisted. For one school, Longshaw Girls', moving between specialist spaces was part of a larger move to new interdisciplinary teaching areas in their new d&t block. Although this was not completed by the end of the fieldwork period, the planning for this had a significant impact on how the teachers concerned viewed the spaces they were using at the time.

Teachers' reactions to these moves towards working across disciplinary spaces can be grouped into four categories: border closing; invasion/colonization; pollution; and border crossing. Each exemplifies an attitude both to the subject of origin and to the space in which it is taught, illuminating how interdisciplinary work impacts upon teachers' ways dealing with and using school spaces, and vice versa.

BORDER CLOSING

A common response to attempts to get teachers to teach d&t in an interdisciplinary way was overtly to retreat behind the borders, both physical and metaphorical, of one's discipline of origin. This was a strategy used with varying degrees of success by a number of individuals and groups, and was particularly employed by male CDT teachers, who invoked health and safety rules as a way of keeping teachers from other specialists out of their workshops. "In the end, I think if you're going to design and make things using resistant materials, you're going to have to come into a workshop sooner or later and use power tools and hand tools. And, unless you're trained to use them and have got experience of using them, you shouldn't really be in charge of children using them" (Simon Guthrie, joint head of technology (d&t teacher), Knype School). Interestingly, though, the same teachers also complained that the new subject, with its increased emphasis on design, required very little in the way of craft skills, and therefore, presumably, the tools needed for it. This suggests that the tools in the workshops, and formal training in their use, had a symbolic function as badges of belonging without which entry was not permitted. This was borne out by the attitude at Turnhill School to Angela Rosen, a textiles teacher, who was required to teach in a workshop, because of a shortage of specialist staff. Angela had originally trained as a primary school teacher and had taught a wide variety of subjects across several phases. At the end of her initial training, in the mid-1970s, she had taken a one-year handicraft course, which covered metalwork, jewelry, and engineering drawing. Fifteen years later this was considered sufficient to qualify her for teaching in the workshops. Given that this department had a history of male resistance to the women from HE doing any work with resistant materials (Paechter & Head, 1996), I assumed that she would concentrate on the metalwork for which she had an official training. Instead, however, she was expected to teach wood-based projects, as did the full-time CDT staff, which supported her in doing this by teaching her basic woodworking techniques. In this case, her handicraft certificate acted as a key to the otherwise closed workshop door; she was allowed in, while another HE teacher, Anita Crawford, who had actually done some work with wood and hand tools within her own area (Paechter & Head, 1996), was kept out:

> "I mean, Angela's actually got training. She's actually been and done some work in jewelry making and in practical work in that area in the workshop

> and she's also qualified, she's qualified to actually teach in a workshop because she's actually gone through the safety procedures and so forth. Anita's in a completely different situation, she's got no…I mean it's just like taking a cover teacher [a non-specialist teacher 'covering' for an absent colleague] and telling them to do practical work which I don't feel is right and we shouldn't do that." (Tom Hardy, CDT teacher, Turnhill School)

Border closing was also attempted by some female HE teachers, though it was less common, and tended to be confined to those who were particularly marginalized in their schools or departments. Rachel McBride, a head of HE who had seen the head of CDT promoted above her to head of technology without advertisement or interview, reacted by drawing away from integrated d&t, although she had been enthusiastic about it at first. She pointed out that work with food involves organizing work areas and ensuring that the rooms are kept spotless, and clearly did not think that people who were not trained in the subject would be able to do this: "People can cook, but it doesn't mean to say that they can teach cookery" (Rachel McBride, HE teacher, Stitchers School).

A third way in which borders were closed involved only one teacher, Simon Guthrie, the joint head of d&t at Knype School, and was a particularly interesting manipulation of curriculum definitions for the preservation of spatial integrity. At this school, as I said above, a shortage of HE teachers in the second year of the study led to a need for CDT staff to contribute to the teaching of textiles, working with students to make drawstring bags. Of the four CDT teachers, three, all men, were asked to do this. Two did so willingly, even eagerly, for reasons, which will be explored below. Simon Guthrie, however, while claiming that he was also teaching textiles, re-drew the borders around that aspect of d&t in such a way that they encompassed things that he felt comfortable about teaching. This allowed him to avoid doing anything remotely related to sewing, with which he clearly felt very uneasy. This enabled him to maintain the spatial borders around him, so that he could remain in his workshop rather than moving to a textiles room at the HE end of the building. At the same time, his reconfiguration of what counted as textiles teaching allowed him to neutralize its (for him) challenging force. This was achieved by defining glass fiber as a textile, and then setting up a construction project using it. Note also the use of the teacher "we" to suggest joint responsibility with students for a decision that was clearly engineered by the teacher to suit the project that he had in mind. "We categorized the uses of textiles in terms of clothing, clothing accessories, construction…and we decided to work in a non-clothing context, so using textiles for something other than just something to wear. And I gave them a design brief to design and make a two and a half metre boat, sailing boat, and race it…I'm trying to show that fabrics are used in car tyres, they're used in all sorts of construction" (Simon Guthrie, joint head of d&t, Knype School). Here, a redefinition of the curriculum allowed Simon to maintain himself within spatial boundaries within which he was comfortable. Curriculum forms were manipulated in the service of the preservation of the spatial status quo.

INVASION/COLONIZATION

Where attempts to close borders were unsuccessful, the consequences could be experienced as an invasion. This was not intended by those moving into others' spaces, but was often how those on the receiving end perceived it. Those invading, while often quite assertive and directive about it, were not deliberately setting out to take over another individual or group's curriculum area. It is nevertheless an apt metaphor, both in terms of the feelings of those who were 'invaded' and in terms of the resultant colonization of some areas of the curriculum by non-specialists in those areas.

In some schools, the introduction of d&t allowed the virtual colonization of one original subject area, and its spaces, by another. This was particularly striking at Bursley School, where there was a completely bloodless takeover of textiles teaching by one female member of the CDT staff, with the result that textiles was increasingly integrated into mainstream d&t while food technology was sidelined. In this school, the HE teachers were fairly powerless compared to those from CDT. The head of HE, who had a very traditional approach, was totally opposed to d&t and had gone on extended sick leave as the research began. A second teacher was also on long-term sick leave, and indeed died during the research period. The remaining HE teacher, Paula Tomson, had become increasingly disenchanted with the subject and was moving into vocational education in the sixth form. This left a considerable hole in the timetable, with HE teaching covered by non-specialist temporary staff, and a greater emphasis on resistant materials within d&t as a whole. There had never really been a lot of textiles teaching done as all the HE teachers were food specialists. When there were insufficient workshops for all the classes, it was suggested that Helen Cartwright, a CDT specialist who had previously worked in the textiles industry, might like to do some textiles work with a group. Initially, Helen shared the textiles room with the head of HE. She used this shared space in such a way as to demonstrate what she saw as her superior knowledge of the subject. "What I was doing is, I had the lesson first and I'd leave all the information and everything on the blackboard, and then she would come in..." (Helen Cartwright, CDT teacher, Bursley School). However, when the head of HE went on sick leave, she took the room and the textiles teaching over entirely, and it became an outpost of the CDT department. As the remaining HE teacher had no interest in textiles, this was unchallenged.

CDT teachers frequently considered textiles a target for colonization, because it could be considered a construction material. At Longshaw Girls' there was also a move, originating outside of the school, to force the full integration of textiles and resistant materials teaching within d&t, explicitly through the manipulation of teaching spaces. At this school, a new integrated technology block was about to be built, and the local education authority (LEA) had designed it in such a way that there would be no separate textiles area; all textiles teaching would take place in multi-materials workshops. The LEA adviser for d&t, Peter Simpson, was quite explicit about his intention to do this through the plans for the new building:

> "well ... you design a building in such a way that people are forced to work in a particular way. And the equipment you supply, again, and the way you set up the areas, forces them to work in a particular way....There's no reason why you can't have sewing machines, knitting machines and so on in a multi-material area." (Peter Simpson, LEA adviser for technology, working with Longshaw Girls' School)

This intention on the part of Peter Simpson was played out alongside a number of power struggles taking place within the department over the degree of integration of d&t. It compounded a growing feeling on the part of the female HE staff that male CDT specialists were invading their spaces. Even before d&t was introduced, the HE team was generally weak. The head teacher regarded it as "the buns department" [field notes], and was keen to push them into integrated d&t. Even within a faculty so under-resourced that everyone relied on using scrap materials donated by industry, textiles stood out as being particularly short of funds. This required considerable ingenuity on the part of the staff. For instance, tearing up used plastic carrier bags and knitting together the resultant strips completed some of the examination pieces prepared by students. Neither of the HE teachers got on well with the recently appointed male head of d&t, who they saw as disorganized and somewhat dictatorial. They had already found that this lack of organization affected their own teaching, and were gradually discovering that this also affected the spaces in which they worked. Consequently they were wary of integrated d&t, which they experienced as a gradual invasion and colonization of their curriculum, spaces and equipment.

In contrast to these two were the head of technology, Tim Winterson, a CDT specialist, and Emlyn Williams, the other male CDT teacher. They were both enthusiastic about integrated d&t, and eager to put it in place. The department had a complicated (and often dysfunctional) system in which groups of students rotated through the various d&t spaces (and consequently through the contributing subjects), more or less every half term. Usually when the groups moved into another area, they also came under the supervision of another teacher. Tim and Emlyn, however, decided that they would like to continue to teach their groups even when they were studying food and textiles. They saw this as an equity issue, as a way of breaking down stereotypes about activities as being suitable for males and females, as well as something that was personally enjoyable for them:

> "But on the other hand, some of the most enjoyable technology I've ever done, personally, has in fact been in textiles. And they learn to sew with me basically because I am determined that man-with-beard will know how to sew because man-with-beard cuts up wood and does mysterious things with machines but in fact he also sews and he also makes teddy bears and all the rest of those things, because it is important that they don't have those sort of preconceptions really." (Emlyn Williams, CDT teacher, Longshaw Girls' School)

The HE staff, who had not been consulted about this decision, resisted these moves by the male CDT teachers to work within the HE areas by attempting to close their borders. They claimed that it was not safe for teachers without appropriate qualifications to teach in food technology areas. "They won't accept that there are health and safety guidelines that [the LEA] are supposed to adhere to, which means unqualified teachers shouldn't be teaching with sewing machines and cookers" (Abigail Simpson, HE teacher, Longshaw Girls' School). Because of their relative powerlessness in the department, however, the two HE teachers were unable to prevent this move into their teaching spaces by the male CDT staff. Their arguments were ignored, and the two men took their classes into cookery rooms and removed sewing machines to their own teaching areas to work with.

Once Peter Simpson's intentions to force the integration of textiles in the new building became clear, these invasive sorties by the male CDT teachers threatened to become a full-scale colonization, with all staff in the department having open access to the textiles equipment in the new multi-material workshops, and the specialist role of textiles teacher disappearing altogether. This led to considerable unease, and deepened the HE staff's feelings of invasion and displacement, particularly as it coincided with the need to vacate their teaching rooms to allow the building work to start.

Gradually, however, the head of technology, Tim Winterson, realized that Lila Sawney, the textiles specialist, was becoming increasingly unhappy, and that this arrangement would not work. While superficially going along with Peter Simpson's plans, he made preparations at least partially to preserve textiles as a separate area:

> "And we're having two multi-material workshops, one technology room, one graphics room, one computer graphics room, a food technology area and a

> resource base on the top floor. Now we couldn't change the plans, but I wanted textiles as well—but I couldn't change the plans because if I'd done that I'd have rocked the boat and put it back another year. So we've got, in the resource area, an area there where I might turn that into the textiles room." (Tim Winterson, head of d&t, Longshaw Girls' School)

Thus, although their initial resistance had failed, by the end of the research period, the HE staff had managed to preserve their subject from complete colonization, though at the expense of continued invasion and the uncertain "ownership" of supposedly shared space in the new technology block.

POLLUTION

The invasion and colonization of HE areas within Longshaw Girls' was bitterly resented throughout the research period. This resentment was expressed through a perception of the male CDT staff moving into the area as an actual and symbolic desecration of teaching spaces and equipment. Because of the ways in which the men used the spaces, the HE staff experienced their presence (and the reminders of their presence that they left behind) as physically polluting. They felt in particular that the CDT men did not treat their teaching areas or equipment with respect, and that they were simply left to pick up the pieces afterwards. "I mean, Abigail and Lila [the HE teachers] have been, well, Abigail voiced to me the sort of feeling that they sort of like, 'oh well, they think it's all so easy. Then they come over here and leave the kitchen in a mess and the sewing machines come back to us broken'" (Maria Christodoulou, CDT teacher, Longshaw Girls' School). For these teachers, the integrity of their teaching spaces had been breached, and the feelings of pollution and pillage that resulted were very clear. "I mean, it costs £70 to have a sewing machine serviced. Well, we'll get through them all soon when they get sent over there and they just send them back broken. You know, and there'll be no money to repair them" (Abigail Simpson, HE teacher, Longshaw Girls' School).

Both Emlyn and Tim were very confident about their ability to work within the food and the textiles areas, seeing it as an extension of their work in the technology of resistant materials. This was also perceived as a lack of respect, but this time it was respect for the specialist skills of the HE staff, expressed through the men's confidence in working in their areas, that was seen as lacking. "I know there's a certain amount of resentment that people feel they can go over to the cookery room and do a bit of cookery, because they're doing cross-curricular stuff. And that's felt as being again, sort of like, you know, 'oh the men feel that they can do these bits of cookery, it's no problem.'" (Maria Christodoulou, CDT teacher, Longshaw Girls' School) This resentment was related to a second way in which pollution is experienced in interdisciplinary work: the metaphorical pollution of the subject area. Previous research (Whitty et al., 1994) has suggested that this is particularly forcefully felt by teachers of high-status subjects, who see the purity of their subject as threatened by encroachment (for example, through imposed cross-curricular themes) by those that are less important in power/knowledge terms. It is clear from my research, however, that teachers of marginal areas also experienced the symbolic pollution of their subjects, not so much through the inclusion of other topics as through the colonizing moves of teachers from other departments. From the point of view of a specialist teacher, once it is accepted that someone without your training can teach your subject, that specialism is itself downgraded and your expertise devalued. Thus the invasion of the food areas and the appropriation of textiles equipment by the male CDT teachers at Longshaw Girls' were experienced simultaneously as an actual and as a metaphorical pollution; the sanctity of their teaching rooms and the purity and status of their subjects were under threat. Consequently, there was an increased emphasis on the skills required to teach in these areas, stressing that it was not as simple as it first appeared. Others could therefore express solidarity with these teachers through emphasizing the high level of skill required to work with students in these rooms and with this equipment. For example, Maria Christodoulou, a CDT teacher at Longshaw Girls' who had close relations with the HE staff, underlined the difference between herself and Emlyn Williams in terms of her appreciation of the high level of these skills.

> "Well I want some INSET [Inservice Education and Training] on using a sewing machine. Because I don't feel confident about a sewing machine. My memories of sewing machines are not being able to thread it. Well, I can thread it but when you get it wrong, and trying to get the bobbin out and doing all the bits, and the needle snaps and you've got to change the needle and all that. I mean, again, Emlyn thinks it's no problem." (Maria Christodoulou, CDT teacher, Longshaw Girls' School)

Maria thus uses her lack of skill symbolically to refuse the colonizer role while simultaneously asking to participate in a formal rite of passage (training) that would allow her to move into textiles teaching in a legitimated manner.

BORDER CROSSING

By contrast, some teachers willingly and co-operatively crossed departmental borders as part of their d&t teaching. This was particularly evident at Knype, where it was further highlighted by the lone refusal of Simon Guthrie, the joint head of department, to do so. The rest of the department were broadly supportive of integrated d&t, to the extent that by the second year of the research they had all, except for one newly qualified member of staff, taught in an area for which they had not originally been trained. In all cases this involved physically moving to a specialist craft area away from their normal teaching spaces, where those who usually worked in that area supported them. At its extremes this meant that boundaries between individual classroom spaces were broken down, so that teachers shared spaces and taught groups between them. Here, Alan McDonald, a CDT teacher, is explaining how he plans to support Martha Hartwell, an HE specialist, when she will be teaching resistant materials to GCSE[2] students:

> "So I'm on at the same time, so I've been actually taking her group up here [i.e. to the resistant materials workshops] so we're next door to each other, and then we go in the workshop and then try and put the two groups together. And then have them in the two rooms so we can move back and forward and actually teach them as a kind of joint group. It might not work cos there's quite a large number, but there's two workshop areas and I'll kind of support Martha and do sort of design work." (Alan McDonald, CDT teacher, Knype School)

The ease with which teachers at this school moved between spaces physically quite distant from each other, and clearly marked as HE or CDT territory, stood out among the research schools. In particular, the appreciation of specialists in one area for the contribution that could be made by those from another was very clearly expressed:

> "I'm just dying to see Alan and John [CDT teachers] on the [sewing] machines and the dyeing, stuff like that. And I think they'll be fine on it because the machine work itself is actually very basic in year 7 and 8. And the design aspect in year 8, the local design work actually lends itself to their skills. So, you know, they teach materials down in the CDT, or the wood, the plastics and the metal, so, you know, it's sort of applying the skills of teaching the materials here as well, only there's different ways of testing it." (Martha Hartwell, HE specialist and joint head of d&t, Knype School)

This generosity was particularly remarkable given that in many ways this department was and remained the most divided of all those studied (Paechter, 2000). The origins of this co-operative approach seem themselves to be both spatial and gendered. At the start of the second year of the study, Jenny Walker, a female CDT teacher, was appointed. She gradually gravitated towards the area of the staff room where Martha Hartwell and another HE teacher, Florence Teleman, habitually sat, and as she developed alliances within the resistant materials end of the department, was joined by John Wallace, another CDT teacher. "We do often sit as a group. It depends very much on where Florence and I are sitting sometimes. Because Jenny will come and sit beside us and then John'll come" (Martha Hartwell, HE specialist and joint head of d&t, Knype School). This social interaction brought the CDT and HE specialists together, so that trust was developed, at least among everyone except the male joint head of department, Simon Guthrie. His increasing isolation within the department in terms of his ideas (he was often at loggerheads with everyone else about departmental policy) was underlined by his physical isolation as the only member of staff not to move between teaching areas. Indeed, it sometimes seemed that the willingness of everyone else to move around was a physical expression of his or her difference from him.

CONCLUSION

Space is therefore a very salient issue in the introduction and execution of interdisciplinary work, especially between subjects for which specialist teaching areas are essential. In vocational and craft subjects, these spaces are both physically and symbolically important and defended as such, border crossing only being permitted or encouraged when trust has already been built by other means. Gender is an important aspect of this. In particular, the ability to defend a subject space seems to be gendered, and to relate to the gender marking of the subject itself more than to the sex of the person teaching it.

In both my studies, the only departments to have their physical and disciplinary borders unwillingly breached were the highly feminized HE specialist areas. CDT teachers were particularly successful at maintaining border controls, although these borders seemed to matter more to male than to female CDT specialists. They used a variety of strategies, and were almost uniformly successful, though sometimes (as in the case of Simon Guthrie at Knype) this was at considerable personal cost. The same strategies used by HE departments, by contrast, did not succeed. This was partly because of the particular circumstances of the departments, and partly because of the way that d&t was perceived both within and outside many schools, as an extension of CDT to include and incorporate at least the textiles aspect of HE (Paechter, 1995a, 2000). At the same time it was clear that male teachers were more likely to continue to move into other people's specialist areas in the face of protest; there were no instances in either study in which a woman moved into another teaching space despite protests from male teachers. At Longshaw Girls' it was the female CDT teacher who refused to cross borders without further training. At Turnhill, where there had been previous attempts on the part of the HE staff to work with resistant materials, these had been confined to their own teaching areas, with no attempt on the part of "untrained" staff to move into the sacred territory of the workshops (Paechter & Head, 1996).

Gendered power relations within and between subject departments in secondary schools are therefore reflected in contestation about spatial relations between subjects and between teachers. This is particularly clearly seen in areas where specialist spaces mark the nature of the subject. For craft and vocational subjects, this skilled use of specialist spaces is fundamental to the self-image of many of the teachers involved. The implications for curriculum change, particularly if it is to involve interdisciplinary work, are clear. Space underpins the curriculum in significant ways, so that changes in curriculum forms can have major implications for how spaces are understood and used. Such changes need to be negotiated with care, particularly if teachers are being asked physically to move from their "home" area to work in another, and especially if there are unequal power relations between amalgamating or co-operating groups. Once resistance has been made concrete through the physical closing of borders, it can be hard to overcome; on the other hand, openness to other forms of knowledge or ways of teaching seems to go alongside openness about shared uses of space (Paechter, 2000). For teachers to work together they have to trust each other and those who have asked them to collaborate; in order to do this, they need to feel secure in the spaces in which they are working. Space matters, in more ways than initially meets the eye.

REFERENCES

DES/Welsh Office. (1990). *Technology in the national curriculum.* London: Her Majesty's Stationery Office.

McGregor, J. (2002). Spatiality and teacher workplace culture: The department as nexus. Paper presented at the ICSEI Conference, Copenhagen. January.

National Curriculum Council. (1990). *Non-statutory guidance: Design and technology capability.* York: National Curriculum Council.

Paechter, C. F. (1993). *Power, knowledge, and the design and technology curriculum.* Unpublished Ph.D, King's College, London.

Paechter, C. (1995a). Subcultural retreat: Negotiating the design and technology curriculum. *British Educational Research Journal, 21*(1), 75–87.

Paechter, C. F. (1995b). *Crossing subject boundaries: The micropolitics of curriculum innovation.* London: HMSO.

Paechter, C. F. (2000). *Changing school subjects: Power, gender and curriculum.* Buckingham: Open University Press.

Paechter, C. F., & Head, J. O. (1996). Power and gender in the staffroom. *British Educational Research Journal, 22*(1), 57–69.

Whitty, G., Rowe, G., & Aggleton, P. (1994). Subjects and themes in the secondary school. *Research Papers in Education, 9*(2), 159–180.

NOTES

1. The first of these was my PhD study, with data collected from 1989–1992, and which focused on power/knowledge issues in the d&t curriculum (Paechter, 1993). The second, funded by the Economic and Social Research Council (award number R000233548) was conducted from 1992–1994, co-directed by myself and John Head and focused on power and gender issues in the negotiation of design and technology. All teachers and schools have been given pseudonyms.
2. General Certificate of Secondary Education, usually taken at age 16. At this time, all students had to take GCSE d&t, with at least half their work in resistant materials. This meant that at Knype, all the HE teachers did some teaching in resistant materials.

CHAPTER 4

SPATIALITY AND TEACHER WORKPLACE CULTURES

The Department as Nexus

Jane McGregor

Although much schooling is clearly spatially constituted, studies of schools tend to take the spatial for granted, either ignoring it completely or interpreting the classroom, staffroom or school as fixed and bounded "modernist spaces of enclosure" (Lankshear et al., 1996). For instance, recent education policy is informed by an almost global discourse of school effectiveness where large-scale surveys de-contextualise schools, reducing them to a series of multivariate factors. While school improvement writing acknowledges the importance of culture(s), this is represented within the container of the institution. Schools are constructed as discrete, uncontested entities existing as circumscribed "spatial and temporal islands." This enables wider social and political problems to be "re-territorialized" as school problems (Nespor, 2002) with teachers, pupils and "the community" easily blamed.

However, schools are increasingly being examined as complex multi-layered institutions through the notion of multiple and embedded contexts that define the workplace and shape teaching and learning (Gutierrerez, 1998; McLaughlin & Talbert, 2001). These are commonly understood and

Space, Curriculum, and Learning, pages 45–59

represented as spatial frameworks organized into a nested hierarchy of different levels or concentric layers (e.g., from classroom to community). However, with notable exceptions such as work at Stanford University (McLaughlin, 1993), the context of teachers' work is under-theorized in significant respects, thereby obscuring as much as it reveals. Drawing on theoretical perspectives from "the spatial turn" in the social sciences (Urry, 1995) and "the cultural turn" in social geography (McDowell, 2000), this chapter argues that interrogating context through spatiality provides a more sophisticated and dynamic understanding of schooling practices.

In Nespor's (1997, p. xiii) conception of institutions as dynamic fluid spaces, schools are not pre-existing bounded entities but "an intersection in social space, a knot in a web of practices that stretch into complex systems beginning and ending outside the school." In this chapter, the subject department in secondary or high schools is explored as an intersection, a node or nexus of relations in that constellation of interactions and practices that continually make and remake the school. In discussing the role of spatiality in teacher workplace culture, space is then neither taken as a fixed physical container for action or as the pre-given social, political or organizational context. Rather than being an arena *within* which social relations take place, space is constituted *through* the social. Space is thus itself enacted (Massey, 1999). This enables us to move beyond a focus on a department as an entity *in* space or the way the spatial influences interactions between teachers, to a view of the department *as* space (-time).

It will be argued that schooling can be seen as a "a web of movements spun from multiple flows of material resources and representations" (Nespor, 1994, p. 11), mobilized through certain densities of interaction—in this case, groupings of department staff. The workplace is then a particular way of configuring space-time, and practice-based theories, particularly network approaches, are likely to be helpful in uncovering the production of spatiality. This challenges a common lack of "spatial literacy" among individuals in institutions (Fisher, 2002), which means there is generally little awareness of the work of producing space through materially embedded practices. It is here that actor network theories may make a useful contribution in tracing the interaction of heterogeneous networks of people, objects, ideas and technologies in enacting space.

It is necessary to outline the notion of place in play here. In the new geography, places are not bounded regions but are contested and shifting, "a grounded intersection" of complex flows and interacting relationships, where socio-spatial practices define and create intersecting places "constituted and maintained by social relations of power and exclusion" (McDowell, 1999, p. 4). A workplace is then composed of heterogeneous and dispersed elements linked together extensively as space-time. Massey (1993, p. 66) argues, "the uniqueness of a place, or a locality ... is constructed out of particular interactions and mutual articulation of social relations, social processes, experiences and understandings, in a situation of co-presence, but where a large proportion of those relations, experiences and understandings are actually constructed on a far larger scale than we happen to define for that moment as the place itself." Of course, social relations and processes sediment into certain patterns, such as the classroom or the staffroom, which reveal the persistence of certain configurations of power relations (Rose, 1999). It is important to remind us that everyday life in schools 'takes place' through iteratively performed and materially embedded practices.

SPATIALITY AND SCHOOLS AS WORKPLACES

As already indicated, the spatial is reciprocally and continually constructed with the social as "a moment in the intersection of configured social relations" (Massey, 1994, p. 265). Space/spatiality is constituted through interactions. "It is the product of the intricacies and the complexities, the interlockings and non-interlockings, of relations from the unimaginably cosmic to the intimately tiny. And precisely because it is the product of relations which are active practices, material and embedded, practices which have to be carried out, space is always in a process of becoming" (Massey, 1999, p. 283). If space is relational, it is never finished or closed. It is not a "completed simultaneity" where all connections are already in existence but is the intersection (or not) of multiple trajectories therefore generative of the new and the possible. Massey (1999) argues for seeing spatiality as *space-time* with places, like spaces, constructed from relations, always in the process of becoming.

As the articulation of relational performances, space produces and is produced by a network of power relations (Rose, 1999). Spatiality in schools reflects the construction and mediation of power relations between different groups, where individuals are differently positioned within a "power-geometry" (Massey, 1993). The very organization of the school imposes rhythms and constraints within a repetitive cycle on all bodies and many school rules and practices are connected to spatiality and embodiment, determining the use of space by students, regulating movement and expected actions. Teachers then draw upon this production of space in demonstrating their authority and maintaining dominance (Shilling, 1991; Gordon et al., 2000). Conversely, space may be used by students to construct their own resistances (Willis, 1977). Space in schools is thus mobilized as a resource in the construction and mediation of power relations including the production and negotiation of gender and sexualities (Coffey & Delamont, 2000).

Space in schools is arguably most powerfully articulated through time as manifested by the timetable. This locates and classifies students (often by age and "ability"), teachers (generally by subject specialist in secondary schools) and "curriculum areas" of knowledge. It is this compartmentalisation that characterizes schools as "loosely coupled systems" (Weik, 1988) made up of apparently bounded, functionally specific spaces such as classrooms and departments that are taken to be unproblemmatically homogeneous. Timetables also reflect competing curricula and values, and may hence be seen as maps of power (Paechter, 1995).

McLaughlin (1993) observed that what teachers do outside the classroom may have as much influence on teaching and learning as what goes on within it. While wanting to disrupt the notions of "inside" and "outside," I focus here on workplace interactions between teachers beyond the classroom. A spatial perspective pays particular attention to the day-to-day social production, reproduction and *meaning* of space. The staffroom, classroom and faculty office are all major locations for teachers' work and association, likewise the fieldtrip, department meeting or more informal occasions such as driving to school. These are all work *places* comprising different practices, with different effects.

THE DEPARTMENT AS A NEXUS OF RELATIONS

Located in a particular "middle ground" of the education system between primary/elementary schools and colleges or universities, secondary schools negotiate tensions between learner- or subject-centred teaching cultures and between the whole school or subject disciplines as the location for the development of professional community (McLaughlin & Talbert, 2001). The subject department emerges as a stable and critical, if under-researched, feature of secondary schools in a variety of national studies (Siskin & Little, 1995). It is also an under-theorized area of research, falling between the large-scale quantitative surveys and qualitative case studies, between the macro scale perspective of the policy landscape and the micro-scale of the everyday experiences of teachers and pupils.

Mainstream interpretations of schools as workplaces are substantially organizational, cultural or micropolitical (Hargreaves, 1994; Ball, 1987; Siskin, 1994). These approaches are frequently juxtaposed as binaries to bring to the foreperceived dichotomies between structural and cultural factors or consensus and conflict (Wallace & Hall, 1997). The boundaries of the department are conventionally defined by the organization and structure, including formal roles and responsibilities and the curriculum expressed through the timetable, "option groupings" and assessment. Naming, architecture and physical proximity are also invoked in delineating a "department." However, theoretical frames relating to administration, the construction of disciplinary knowledge and the professional identities of teachers taken separately are inadequate in tracing this locus of intersecting influences, activities and knowledge. In a spatial formulation they are fundamental and socially constructed locations for teachers' work and identity, which are heterogeneous "dynamic structures, challenged in subtle ways everyday" (Gutierrez, 1998).

In addition, as a "core teaching context" (Talbert, 1995, p. 89), there is a growing interest in the department as a crucial arena for workplace collaboration and learning, as a mediator of school level and national "external" influences and policies. Departments are thus increasingly identified as the "missing link" in school improvement (Harris & Bennett, 2001). While broad forces influence the scope and organization of the school curriculum, departments generally have a high degree of control over course construction, employment of staff and assignment to groups. Through immediate everyday interactions they therefore continually shape what it means to be a teacher in a particular department and hence influence students' experience in the classroom.

As illustrated by various empirical studies, the department is a major location and primary reference group for teacher interaction with a variety of forms and content of association (McGregor, 2000; Siskin & Little, 1995). The nature and discourse of the subject "matter" provides a focus for teacher identity often relating back to university education (Reid & Caldwell, 1997). This influences the way teachers conceptualize and teach "subjects" and interpret and put into practice generic policies. Distinctive subject subcultures are thus created and maintained through an intersection of teacher identity, socialization and status within a hierarchy of subjects/knowledge (Paechter, 2000). The culture is manifested through agreed practices and policies, artefacts and displays, rituals, relationships and language. Subject departments are not simply bounded, homogeneous units therefore, but potentially dynamic sites for the continual negotiation and reframing of subject knowledge and pedagogy.

Conventionally, subject classrooms are architecturally grouped in a school and teachers may spend significant and substantive time with colleagues assigned to the department, particularly in department offices: 'for these teachers, private quarters are in communal and public space' (Siskin, 1991, p. 142). Departments may provide teachers with the professional support, trust and openness, which encourage the synergy of strong forms of collaborative joint work where new learning and knowledge are created (Fielding, 1997). However, they may also be self-referential, exclusive and competitive enclaves of professional self-interest, as described by Hargreaves (1994) in his depiction of "cosy collaboration" and Balkanized teacher cultures. A more active conceptualization of the department may be reached though the deployment of practice-based theories such as Actor Network Theor(ies) and situated learning theory. These move away from a structural analysis of behaviour in terms of pre-existing systems and contents to a focus on actors' contributions to the social order. There is a greater emphasis on emergent processes and the roles of association and negotiation in constructing the social.

The model of communities of practice as developed by Lave and Wenger (1999) derives from the notion of situated learning, as participation *in* learning rather than *through* acquisition. Knowledge is created through practice, so learning is social and participative rather than internally cognitive. This implies relational rather than zero-sum understandings of power as the influence shaping action (Sharp et al., 2000). Knowledge is not some*thing* that can be carried around or given away but is located *in* the situation, or rather in terms of spatiality, *as* an element in the situation. If members of a group through the making of meaning create new practice together, this reciprocally brings into existence the grouping or community. The spatiality of the practices in which members participate is a particular "practice-relevant configuration" (Nespor, 1994) which defines the possibilities for learning to "take place" (or not). However, despite modifications, Wenger's model of communities of practice (2002) lacks a developed spatial perspective and correspondingly an analytic of power. If power is seen as a (potentially positive) constellation of relations (Paechter, 2000), then interactions and learning in communities of practice result from lateral modalities of power such as negotiation and collaboration, which are more likely to occur in particular (space-time) locations than others.

LOCATING WORKPLACE INTERACTIONS—THE EMPRICIAL STUDY

In exploring patterns of association such as collaboration, and the creation of particular spatialities, a methodology was developed which asked staff in two secondary schools serving 11–18 year olds in semi-rural areas, to indicate on a grid where, when and with whom they commonly interacted (McGregor, 2000). The department proved to be the major location for associations of all kinds with over 42 percent of the total reported (McGregor, 2003). In both schools, the importance of the department office was emphasized, confirming it as the major *place* for interaction of all kinds, whether explicitly social talk or strong forms of collaborative work such as joint planning in department meetings.

At one of the schools, Kingbourn, the staff generally characterized themselves as working collaboratively but, although the majority met in the main staff room at morning break, they usually sat with their departmental colleagues in clearly defined locales. The larger and more successful school in examination outcomes, Kingbourn was a Training School for new teachers. Interactions with "trainees" based in departments were indicated to be important, particularly in relation to some of those stronger forms of joint work, for instance in mutual observation with feedback or designing and preparing lessons together.

In order to elicit the subjective meanings teachers attached to work spaces, volunteers photographed or created drawings or mental maps of their workplace. Departments were frequently shown as "home," and associated with positive feelings through symbols such as hearts, sighs of contentment or inclusion in a "happy net." This echoes the finding that "at their strongest, subject departments form a powerful intellectual, social and political home for teachers" (Siskin & Little, 1995, p. 185). However, as Siskin (1991) points out, a house is not necessarily a home and some teachers in the study perceived themselves in very different "places" to the colleagues with whom they were in close physical proximity.

The Science Department at Kingbourn

The Science curriculum area at Kingbourn (comprising chemistry, biology, physics, geology and agriculture/rural studies) was located in the relatively favourable position of being part of the core curriculum in compulsory schooling, thereby having significant advantages in time and resource allocation. In years 10 and 11 there were the alternative routes of single or double science. For the sixth form, A levels were offered in physics, chemistry, biology and geology with a GNVQ Integrated science. Student examination performances in science at all stages were well above the national averages. The specialist laboratories were clustered in a block abutting a farm and surrounding the technicians' "prep room." The Head of Department, Gregg Parfitt suggested that the physical proximity of the laboratories and the teachers working in them was important, as it enabled innovations such as CASE (Cognitive Acceleration in Science Education) "to enter and be embedded in the classroom."

The Science curriculum area or department was viewed in the school as cohesive and effective, but rather separate from the whole staff. This situation was embodied by attitudes to the use of the small room variously described as the Science office, marking or work room. "One of our most successful departments if not the most successful department in recent years certainly the largest has the smallest amount of space to work collaboratively. That's the science department" (Deputy Headteacher). Gregg Parfitt was clear that "*the physical space*" of the room was "one of our greatest strengths." He felt that it was important for him not to have his own separate office as this would have the effect of "decreasing open dialogue dramatically." The space was not uncontested however, as the Headteacher, a powerful advocate of cross-curricular work consciously manipulated space-time to facilitate whole-school collaborative working through the creation of enquiry groups supported through timetabled meetings, the remodelling of the staffroom, and the disposal of some department offices. The relationship between the Science department, the Head and the rest of the school was a microcosm of the conflicts highlighted in the educational literature over the appropriate focus for the development of professional community and school improvement. Until recently, the small, cramped Science office had not been extended. It was suggested by the Bursar (responsible for refurbishment) that the school improvement project had contributed to a wider range of interactions "within" departments:

> "There became this developing body of researchers that people, actually teachers, shared a lot of information and felt free to talk about teaching. It wasn't just social time when they were in their department, they were actually talking about teaching and it was a really strong way of developing teacher knowledge and confidence about teaching and learning. And so we gradually eased up a little bit and developed, not social areas for staff, but work areas for staff where they could be social, where they didn't have to come across here (staffroom) at break time or lunch time and work."

The staff in Science were a mix of male, female, roles and responsibilities, age and experience, including two previous Heads of department. As in many such schools, teachers had a variety of roles and responsibilities, so the Head of Department was a member of the Year team led by another science teacher. The apparently fluid power relations were perceived generally as co-operative, rather than a hierarchical power-geometry, and did not appear particularly role or length of experience based. For example, a newly qualified teacher and a technician as part of a validating process checked completed schemes of work. "(We are) collaborative, co-operative, not hierarchical. Collaboration means having leadership across a wide range of staff, you don't have to be an old hand in the department to have a leadership role. It encourages people to develop themselves" (Gregg Parfitt).

In contrast to some schools, the five technicians were integrated socially into the department as well as through their support function (and one has since become a teacher in the department). The Curriculum Area Leader played a pivotal role in creating the conditions for a collaborative "community of practice" within the Science department. A talented and committed teacher, Gregg Parfitt had been offered more senior jobs within the school but was adamant that he wished to retain the head of department role. He believed it to be a crucial position for introducing and evaluating ideas in the classroom.

As indicated earlier, interactions create spatiality and in this case associations and ways of working together were deliberately facilitated by Gregg and deputy Delphine to enhance strong forms of collaborative work, such as joint planning and the collective testing of ideas. The choice and adoption of the GCSE combined modular examination course had been a powerful force in bringing the staff together "in science" to break down the traditional subject boundaries of physics, biology and chemistry, by discussing and "re-placing" subject content and pedagogy. The modular course was seen to be fundamental in bringing disparate subject staff together, as students are passed from teacher to teacher within a structured course that demonstrates progression, partly through regular assessments. It was necessary for staff to work closely in sharing planning, timing and knowledge of the cohort of students. It could be argued that it is also a form of surveillance whereby the result of their teaching, marking, etc. becomes more visible and open to scrutiny during the school year.

Most courses were created collaboratively in review and planning meetings during the summer term where teachers discussed and shared understandings about scientific concepts, assumptions about appropriate pedagogy and desired outcomes, and forged common agreements often through lively debate and negotiation. These clearly demonstrated the kind of "critical practices of adaptability" or strong forms of joint work that Little (1982) identified. In such meetings, a range of modalities of the operation of power could be observed, as individuals debated grounded interpretations of the syllabus and volubly lobbied for particular approaches. Negotiation, collaboration and "seduction" seemed to be frequently employed; rarely domination or imposition (Allen, 1999) and the spatialities of these meetings were qualitatively different from other formal meetings or informal times. These types of interaction were used to explore, agree and fix a collective orientation and new artefacts such as work-schemes were generated, reifying the collaborative outcomes and arguably creating "immutable mobiles" (Latour, 1997), which could translate the knowledge content and practice between classrooms.

Despite the pressure of work, new methods and ideas such as CASE were welcomed as a deliberate opportunity to collaborate and learn together at a similar level of competence (while at the other school they tended to be seen as an unwelcome imposition). Trust, humour and mutual respect were all cited as contributing to the coherence of the department, the knotting of social and professional relations, which allowed staff to share successes and failures and to acknowledge their own learning. This was modelled particularly by Gregg and Delphine. Thus a department culture of openness was created and sustained in an ongoing fashion, manifested particularly by the spatiality of the department office.

Over the course of break times the majority of the 25 staff visited the tiny science office, providing the opportunity for casual, serendipitous contact as well as more focused social or work-related conversations. "A huge number of interactions are going on, an assemblage of social talk and humour, interchanges about work schemes, lessons and resources, also students (but noticeably not much storytelling). There is a lot of talk about the curriculum ...Talk goes seamlessly between personal to education/curriculum issues" (Fieldnotes, Kingbourn). Such everyday interactions create the contexts and interpret the contingencies out of which the next action springs. This was located particularly in shared talk of different kinds, with the interlacing of work-related and social talk arguably crucial in constructing the "community of practice." For this department, their "common ground" was the teaching of science, which many of them cared about passionately, and their practice was articulated through a common language, which was often quite inaccessible to "outsiders."

The Science department was a network that included technologies and objects as active participants, determining solutions, circulating ideas and affecting actions (Law, 1992). The science office and the "prep room" were the repository for information and artefacts that teachers needed on a daily and sometimes hourly basis. For example, in the "prep room" there were the sheets requisitioning essential lesson equipment to be provided by the technicians and the lists of public examination times. In the office there was a filing cabinet of student records, which teachers referred to and filled in particularly at the end and beginning of modules, and a computer, which was networked and regularly used to store and call up schemes of work. On the walls were timetables and photographs of the students in the module groups. Thus teachers were drawn to the room to refer to these objects which were very much part of the ongoing conversations between them. Here we see the network of relations as heterogeneous, incorporating objects that exerted their own agency in relation to the performance of space that was the department. Hence the photographs of groups who might be quite new to a teacher attracted people to the room

In addition to the official documents, there were collections of photographs on the walls, showing members of the department at past celebratory events, in relaxed or humorous poses. These displays produced a space-time stretched out beyond the limits of the room, articulating a sense of the department as social, even familial, as well as professional, and of having a shared and valued history. The workspace was therefore constructed as more porous and dynamic, not simply in one direction of work impinging on home. Observation of the language and interchange at breaktimes reinforced this idea, in the assemblage of story telling and dialogue where talk about families was interwoven with discussions about the curriculum. The Science department office was therefore much more than a context in which interaction occurred. Social meanings were embedded within it that served to facilitate certain types of interaction *constructing* the space. This comprised a practice-relevant configuration, a particular juxtaposition of people and artefacts that enabled and focused the interaction including wider discourses drawn on from "beyond" the department. This was part of the ongoing creation of the department as a dynamic enterprise, a network of formal and informal relations and a locus of wider socio-spatial flows.

With the exception of one physics student teacher, all the teachers belonging to the curriculum area identified themselves firstly as "Science staff," suggesting an identity evolving as a function of ongoing interchanges and practice. However, most subsequently identified themselves as subject specialists supported by the department in working "outside" their specialisms. The mentoring of teaching students or trainees was also a major element in the intersection of identity, organizational and subject affiliation. "The whole department want to progress so much and know they have to do it together. As students coming in we are welcomed and supported by the department. We are seen as having new and fresh ideas" (Student teacher). The induction of students and new members of staff "into the department" included the formal process of providing handbooks, structured observation and mentoring sessions. However, the everyday space of the department office was observably an important dimension of induction. Student teachers were able to take part in social and practice-related interchanges as members of the group, thus exemplifying "legitimate peripheral participation" (Lave & Wenger, 1999). By working in the office when marking assignments or preparing lessons, their learning about how to be a science teacher was supported through immediate (and relatively unthreatening) access to comment, advice and humour.

Others saw the science staff as a particularly cohesive group. However, although they indicated a large number of interactions, and, in interview, emphasized their interdependence, they produced a significantly lower percentage of interactions "within" the department (36%), with more than usual links reported "beyond." These included the cross-curricular groupings, the Science Association and research groups within and beyond the school. Gregg was well aware of this: "(The Head) has been cutting about the issue that Science is separate—we are relatively separate in that we don't come over (to the main staffroom) at break—we *are* integrated, more integrated ironically than other departments." Here we may see strength in links and inter-relations where the department was actually functioning as a more open system. Practice was not simply the observable actions of individuals, but also distributed across space-time as teachers drew on their experience and education and that of others, as well as mobilizing textbooks, videos and the Internet.

CONCLUSION

In this chapter, schools as institutions are taken to be "precarious geographical achievements" (Philo & Parr, 2000) made and remade by configuring practices and relations. As particular workplaces, they are constellations of relations with their uniqueness constructed from the juxtaposition of relations connected in/as space-time to wider flows. Within this network of relations, "the department" may be seen as a particular locus of such (power) relations, made and remade by the interaction of people, objects, ideas and discourses. This nexus of inter-relations comprises forces and influences that intersect and interplay with organizational and curricular structures and histories, social and political constructions and the material environment in which people experience their workplace. The Science department at Kingbourn was one such configuration, a "space-contingent field of practice" (Nespor, 1994).

Approaching the workplace as spatially constructed thus moves beyond the concept of multiple contexts to suggest a topological approach (Mol & Law, 1999; Nespor, 2000) to understanding what takes place in schools. Practice-based theory also suggests recognizing places as co-constitutive of knowledge, practice and identity with practice-relevant configurations as the pre-requisite actor-networks, which may comprise a community of practice. Such heterogeneous networks thus accomplish events and make sense of them in particular sites. In the case of the Science department at Kingbourn, the laboratories, procedures for ordering chemicals, work-schemes created collaboratively and stored on computers, and so on, interacted with the continually enacted social and professional relations to create what was described as "a very satisfying place to be in" (Bella Miston, Kingbourn). Thus we may move conceptually away from the notion of pre-existing performance teams or communities of practice into which individuals move, but rather suggest that the department is performed through the various constitutive activities and thus is itself a way of configuring space-time.

Space is therefore not simply the untidy backdrop to teacher and student interactions, but filled with professional and political significance. Seeing space as actively constructed and performed requires greater consideration of relations of power. Work needs to be done on spatializing models such as communities of practice to allow a more sophisticated understanding of the operation of power through lateral modalities such as collaboration. This in turn may inform current discussions about the nature of leadership in schools. If power is immanent rather than hierarchical, with particular assemblages of space integral to organizational forms, a spatial analytic of power and how everyday relations interact and intersect with wider flows, particularly through network space, provides an alternative view to current unhelpful binaries such as structure and culture or micro and micro perspectives.

Where the organizational unit of the department intersects with collaborative cultures or communities of practice there can be a powerful synergy, creating new possibilities for learning of all kinds. As a way of organizing time and space, the department as a work *place* reassembles the conditions necessary to be a science teacher. But further, when *knowing* is also seen as distributed, teaching may be seen not just as the product of individual teachers working individually in isolated classrooms, but as "a collective accomplishment of groups of teachers working together" (Nespor, 1997, p. 12), thus opening up other spaces for learning and action. Through the intrinsic openness of space-time, we see the possibilities for making spaces of association for adults and students in schools that are generative of mutual learning and understanding.

REFERENCES

Allen, J., (1999). Spatial assemblages of power: From domination to empowerment. In D. Massey, J. Allen & P. Sarre (Eds.), *Human Geography Today*. Cambridge: Polity Press.

Ball, S. (1987). *The Micropolitics of the school: Towards a theory of school organization.* London: Methuen.

Coffey, A., & Delamont, S. (2000). *Feminism and the classroom teacher: Research, praxis and pedagogy*. London: Routledge Falmer.

Fielding, M. (1997). On collegiality. Paper presented at the *European Conference on Educational Research.* Johann Wolfgang Goethe-Universitat, Frankfurt am Main Germany.

Fisher, K. (2002). Re-voicing the classroom. Unpublished PhD thesis, Adelaide, Australia.

Gordon, T., Holland, J., & Lahelma, E. (Eds.), (2000). *Making spaces: Citizenship and difference in schools.* Basingstoke: Macmillan.

Gutierrez, R. (1998). Departments as contexts for understanding and reforming secondary teachers' work: Continuing the dialogue. *Journal of Curriculum Studies, 30*(1), 95–103.

Hargreaves, A. (1994). *Changing teachers, changing times: Teachers' work and culture in the post-modern age.* London: Cassell.

Harris, A., & Bennett, N. (Eds.), (2001). *School effectiveness and school improvement: Alternative perspectives.* London: Continuum.

Lankshear, C., Peters, M., & Knobel, M. (1996). Critical pedagogy and cyberspace. In H. Giroux, C. Lankshear, P. McLaren, & M. Peters (Eds.), *Counternarratives cultural studies and critical pedagogies in postmodern spaces.* New York: Routledge.

Latour, B. (1997). On actor-network theory a few clarifications. University of Keele: Centre for Social theory and Technology .

Lave, J., & Wenger, E. (1999). Legitimate peripheral participation in communities of practice. In R. McCormick & C. Paechter (Eds.), *Learning and knowledge.* London: Sage.

Law, J. (1992). Notes on the theory of the actor network: Ordering, strategy and heterogeneity. *Systems Practice, 5*(4), 379–393.

Little, J. W. (1982). Norms of collegiality and experimentation: Workplace conditions of school success. *American Education Research Journal, 19*(3).

Little, J.W. (1990). Conditions of professional development in secondary schools. In M.W. McLaughlin, J. Talbert, & N. Bascia (Eds.), *The contexts of teaching in secondary schools: Teachers' realities.* New York: Teachers' College Press.

Massey, D. (1993). Power-geometry and a progressive sense of place. In J. Bird, B. Curtis, T. Putnam, G. Robertson, & L. Tucker (Eds.), *Mapping the futures: Local cultures, global change.* London: Routledge.

Massey, D. (1994). *Space, place and gender.* Cambridge: Polity Press.

Massey, D. (1999). Spaces of politics. In D. Massey, J. Allen, & P. Sarre (Eds.), *Human geography today.* Cambridge: Polity Press.

McDowell, L. (1999). *Gender, identity and place. Understanding feminist geographies.* Cambridge: Polity Press.

McDowell, L., & Sharp, J. (Eds.). (1999). *A feminist glossary of human geography.* London: Arnold.

McGregor, J. (2000). Travelling together: Teachers meeting the challenge of collaboration. *Forum, 42*(1), 15–23.

McGregor, J. (2003). Collaboration in communities of practice. In N. Bennett & L. Anderson (Eds.), Rethinking e*ducational leadership: Challenging the conventions.* London: Sage.

McLaughlin, M. (1993). What matters most in teachers' workplace context? In J. Little & M. McLaughlin (Eds.), *Teachers' work: Individuals , colleagues and contexts.* New York: Teachers' College Press.

McLaughlin, M., & Talbert, J. (2001). *Professional communities and the work of high school teaching.* London: University of Chicago Press.

Mol, A.M., & Law, J. (1994). Regions, networks and fluids: Anaemia and social topology. *Social Studies of Science, 24,* 641–671.

Nespor, J. (1994). *Knowledge in motion. space, time and curriculum in undergraduate physics and management.* London: Falmer.

Nespor, J. (1997). *Tangled up in school: Politics, space, bodies and signs in the educational process.* New Jersey: Lawrence Erlbaum.

Nespor, J. (2000). Topologies of masculinity. Gendered spatialities of preadolescent boys." In N. Lesko (Ed.), *Masculinities at school.* Thousand Oaks, California: Sage.

Nespor, J. (2002). Studying the spatialities of schooling: Book review of making spaces: Citizenship and difference in schools. *Pedagogy, Culture and Society, 10*(3), 483–492.

Paechter, C. (1995). *Crossing subject boundaries. The micropolitics of curriculum innovation.* London: HMSO.

Paechter, C. (2000). *Changing school subjects.* Buckingham: Open University Press.

Philo, C., & Parr, H. (2000). Editorial institutional geographies: Introductory remarks. *Geoforum, 31,* 513–521.

Reid, I., & Caldwell, J. (1977). Why did secondary PGCE students choose teaching as a career? *Research in Education, 58,* 46–58.

Rose, G. (1999). Performing space. In D. Massey, J. Allen, & P. Sarre. (Eds.), *Human geography today.* Cambridge:Polity Press.

Sharp, J., Routledge, P., Philo, C., & Paddison, R. (Eds.), (2000). *Entanglements of power: geographies of domination and resistance.* London: Routledge.

Shilling, C. (1991). Social space, gender inequalities and educational differentiation. *British Journal of Sociology of Education, 12*(1), 23–44.

Siskin, L. (1994). *Realms of knowledge: Academic departments in secondary schools.* London: Falmer Press.

Siskin, L.S., & Little, J.W. (Eds.). (1995). *The subjects in question: Departmental organisation and the high school.* New York: Teachers College Press.

Urry, J. (1995). *Consuming places.* London: Routledge.

Wallace, M., & Hall, V. (1997). Towards a cultural and political perspective. In A. Harris, N. Bennett, & M. Preedy (Eds.), *Organizational effectiveness and improvement in education.* Buckingham: Open University Press.

Weik, K. (1988). Educational organizations as loosely coupled systems. In A. Westoby (Ed.) *Culture and power in educational organizations.* Milton Keynes: Open University Press.

Wenger, E., McDermott, R., & Snyder, W. (2002). *Cultivating communities of practice.* Boston: Harvard Business School Publishing.

Willis, P. (1977). *Learning to labour.* Farnborough: Saxon House.

CHAPTER 5

LEARNING, PARTICIPATION AND IDENTIFICATION THROUGH SCHOOL GROUNDS DEVELOPMENT

Greg Mannion

INTRODUCTION

The focus of the case study described in this chapter was children's experience of participation and learning through school grounds development in Scotland. The research was conducted shortly after the shootings in Dunblane primary school (see Cullen, 1996), which provoked a rise in school security measures and probably added to the general rise in fears among parents for their children's safety (Valentine, 1997). Alongside this, commentators have expressed fear for the loss of children's "public space" by curfews and surveillance (Valentine & McKendrick, 1997). This drive to protect children from danger contrasts with popular concerns for the environment, active citizenship and the social inclusion agenda. These current policy contexts add different layers of significance to the case study about children's participation.

Space, Curriculum, and Learning, pages 61–78

The analysis of the cases suggests that we must take greater cognisance of the emergent, spatial and material nature of the learning event. Looking at school grounds developments makes us reconsider the importance of how people, places and learning inter-relate. While traditionally, schools grounds have seemed similar to prison exercise yards, once grounds are reconfigured, new values are asserted. The informal, incidental and seasonal aspects of places and learning are recovered with many new sites expressing more child-friendly, familial and collective values. Places, material objects and engagement in matters that make a difference are seen to be the important ingredients in learning.

The chapter describes some of the participatory elements of the research approach (Anderson, 2000) and the efforts made to conduct the enquiry within an ethic of care and authenticity. The approach was responsive to calls for greater participation by the 'researched' in research processes and to the potential for children to be active agents and social actors in their own right (James & Prout, 1990). What follows is a local, collaborative image-text designed to perform a visual sociology of children's learning through school grounds change processes. The inclusion of photographs is central to exploring how identity, space and learning can be framed in a connected way. In order to more openly acknowledge the constructed nature of photography I foreground the relational aspects of the research process and the effects of this text. The text adopts a performative approach (Edwards et al., 2002) where all research "findings" are better seen as proactive agents in the co-construction of reality rather than as depictions of an already "discovered" world (Harper, 1998). Before the analysis however, there is a need to put forward some key theoretical positions that underpinned the approach. I therefore first set out the pertinent theoretical understandings of identity formation, place and learning.

DISLOCATING PLACE, DE-CENTRING SELF AND ACTIVE KNOWING

Conventionally, places are represented using two-dimensional maps. More recently, geographers and sociologists have begun to argue that places are in fact always changing and are comprised of human and non-human relations cut through by forces and ideas from other times and places. The spatial turn in sociology and the cultural turn in geography both play on the idea that space is socially constructed, unstable and ever emergent. The events that unfold in a place cannot be easily distinguished from the changing nature of a place. Many now agree that in fact places are more like 'events' that cannot be fully captured by traditional mapping. Massey (1994, p. 115, italics added) asks us to think of "place" as "a particular part of, a particular *moment* in, the global network of ...social relations and understandings." As a result, places (such as home and country) that we once considered as stable and fixed have become unstuck and fluid. The events that we may have once considered occurred *in* time and space are in fact the sorts of practices that continually *constitute* time-space.

This non-foundational reading of place brings with it a feeling of dislocation or a sense of loss of stable ground upon which we once "stood," both metaphorically and in material terms. The consequences for our identities are far-reaching. If places are not stable in their essence, then the prospect of achieving a stable identity is also undercut. Indeed, many authors invite us to consider less fixed notions of the self. Mouffe (1998) suggests that we are not single rational individuals but "we are in fact always multiple and contradictory subjects," inhabitants of a diversity of communities. We may be as multiple as the social relations in which we participate and the "selves" they define. Hall (1996) contends that, like the places we inhabit, our identities are not fixed or unified. In late modern times, we are becoming increasingly fragmented and fractured. We construct multiple identifications "across different, often intersecting and antagonistic, discourses, practices and positions" (Hall, 1996, p. 4). A recurring effect of these fluid theories of places and selves is a challenging of foundational views—in fact a dismissal of what might once have been considered as essences. This is achieved via a common strategy of shifting from the use of nouns (such as place and identity) to verbs (spatializisation, production of space, identification) (Mannion, 2002). Objects are only invariable to the degree that they hold themselves constantly in fluid form. These theoretical positions with respect to place and identity lead us to a central insight. Places and people are not fixed entities with essences; but are ever-emerging processes under production and are always the subject of power and politics.

The shift from the noun to the verb is also apparent in theories of learning. Some theorists argue that knowing is a process that is continually mediated, situated, provisional, pragmatic and contested. Situated theorists (e.g., Wenger, 1998) see learning as part of a process of engagement in a community of practice where expert members initiate "new-comers" over time. Fenwick (2001) draws on evolutionary and systems theory to argue that we need to understand learning as part of the systemic way in which people and environments "co-emerge." The view here is that knowing cannot be understood except in terms of how human and non-human systems connect. This ecological understanding asks us to become aware "of how one exists simultaneously in and across these levels, and of how part and whole co-emerge and co-specify one another" (Davis & Sumara, 1997). The co-emergent and biological perspective is different from the situated learning position by being less anthropocentric. Instead, knowledge, identities, and environments emerge in a coupled manner *together.* This eco-social view of knowing is founded on the premise that "cognition exists in the interstices of a complex ecology or organismic relationality" (Davis & Sumara, 1997). Knowing, then, continually 'spills over' within ever-emergent contexts. Knowledge is not a 'thing' that can be taught but flows verb-like within relationships between people, places and a range of other features of systems, which are they changing.

RESEARCH PERFORMANCES

The connected theorizing of identity, place and learning presents a strong challenge to the practice of realist sociology. How do we conduct an empirical sociology of learning when these time-space events appear to be so slippery and emergent? More importantly, can we still capture the effects of power in our research given its embedded nature? Thrift (2000) argues that the very nature of an event denies us opportunity for representational realism. Because entities are emergent and cannot be represented "once-and-for-all," we need to adopt non-representational modes of communication. Radical proponents of this emergent position suggest we cannot divide knowledge from knowers at any time, that there is no "out there" separate from an "in here." In research terms, our interpretations of the action cannot be dissociated from our embeddedness in *on-going* action. While events are not easily replicable, the effects of events (sensory, emotional, rational, political and so on) might be better "performed", rehearsed or improvised for readers of research texts.

In response to these epistemological problems, the chosen "methodology" for my research was, in part at least, a performative one (Newman & Holzman, 1997; Law & Urry, 2002). Because understanding is always an unfolding process (uniting thinking, feeling, speaking and acting at a time and in a place), readers will play a part in determining how this works out. As a performance, it may succeed in being evocative of the effects of power or it may not. The text is staged as a different kind of research performance. The role of the reader in this is to engage in a new activity where things are not to be "known" in the habitual manner but to be engaged in as a new practice. The step-change is from the desire for static knowledge to the enactment of "knowing performances." Things (nouns such as "person," "place" and "event") are unknowable because they are so verb-like. Verb-like things are better *performed* if we are to engage with them (Newman & Holzman, 1997). The shift is from epistemology (where what is known depends on perspective) to ontology (what is known is also being made differently).

Figure 5.1. The den. The children guide me to a site of social significance for some of the girls under a line of bushes. They demonstrate how the place is regularly rearranged to suit their changing purposes. This site and was seen by the pupils as being an on-going grounds development project that had been instigated and carried out by children in the main.

Researchers of the "emergent" need to find ways of allowing their on-going lived character of person-place events to "spill over" so that they can be experienced by others. One tool that is available to the researcher of the "emergent" is photography. Photographs have often been used as uncomplicated documents in support of realist tales. However, with the advance of digital technology in particular, there is a widespread and fundamental challenge to the assumption that photographs correlate with reality in an unmediated manner (Denzin, 1997). For some time, anthropologists have seen photographs as both creations and reflections of what is visible within the scope of a lens. They are seen to frame and reflect the relationship between photographer and subject where both play roles in shaping the image (Collier, 2001). In other words, the subject, viewer and photographer all play vital roles. Reading photographs can be a verb-like, fluid event. Photographs allow readers to vicariously experience happenings and draw their own conclusions. They can help us reflexively analyse the complexities of the elusive events that unfolded in the field. Viewed in this way, photographs can be used to respond to the epistemological challenge of researching learning within person-place-events. This rests on the idea that the meaning of the photographs presented herein were and are jointly constructed by both the makers (me, the children, and the bits and pieces of natural and non-natural material) and you, the viewers who may adopt different positions.

Photo-texts are useful to researchers because they can be performative sites for knowledge generation. The interaction between images and written text can enable an exposition of some global and local forces that were influential. This image-based approach seeks to include the fleeting, distributed, multiple, sensory, emotional, and kinaesthetic through the inclusion of person-place events. In dramatic terms, I hope to provide multiple points of view, rather like being able to see backstage, front-of-stage while reading the critics review. We should resist seeing these photographs as simply lies or as truths. Instead, we should attend to the kinds of truths and the kinds of lies they attempt to tell (Wagner, 1979). Research texts as performances seek to bring into being a new form of life because person-place interaction patterns are continually emerging and cannot be captured "once and for all."

RESEARCHING SCHOOL GROUNDS

The research took a two-phased, case study approach, which was contextually well defined and field-oriented (Stake, 1995). The first phase involved intensive fieldwork in a playground in one school. The choice of school was dependent not on it's typicality but on its potential for being instrumental in providing insight into the issue of children's participation and the spatial aspects of participation and learning in a school grounds setting. The choice was also made on the probability of imminent change to this particular school grounds, the school's location, staff's willingness to participate in the research, and my perception that there was capacity and desire for children's involvement in the processes of change. Regular visits to this school were made (on average once a week over a two year period) and considerable time was spent in the playground as part of almost every visit. Phase one involved making a visual audit using still photography of children's day-to-day use of the playground before changes occurred. Other methods used included two-dimensional maps, interviews, and guided tours provided by children.

On introducing myself to the pupils in classrooms, I usually described my role as researcher/story teller of children's use of play areas. I gave more details about my role depending on time available, the age of the children and the interest shown by the participants. I showed children my camera in advance of my interactions with children during fieldwork. I told them I would not be using to take photographs of children "who particularly wanted their photograph to be taken." Neither would I be using it "secretly" as a spy. Instead, the aim was to collaborate with the children in the making of photographs in an open, performative way. I told them that children who were already engaged in playground activity would be asked to continue with their activity or sometimes "re-perform" the same activity for the camera. Similarly, children often provided me with a demonstration of critical aspects of their activities in key sites and I would attempt to record this on camera. I was careful to describe my role to children as "other-than-teacher." I was careful not to be seen to report stories back to staff or to reprimand children if they were doing something "wrong." I told children I was not in the business of particularly looking out for dangerous incidents or bullying, though I did intervene if I was the only available adult and some child's safety was being compromised.

Over a pilot period, I made subtle changes to how I approached photograph production. As I got to know the children, I discovered how best to invite them to jointly co-construct shots for the camera without ruining the "flow" of their activity. My interest areas were progressively focused, as I became aware of activity types, places and age groups that I had not recorded at all. Taking photographs in this manner was not naturalistic. My research approach involved a distinct intervention in children's participation in the play space, involved children's collaboration as informants and performers. Over two years of intermittent fieldwork, I took 192 photographs. I became actively involved in one or two school grounds projects on-site. This necessitated the adoption of a form of action-researcher role at times, which required a more reflexive approach to researching participation in design and implementation of change. My approach sought to be sensitive to the diverse needs of children, staff and others, while trying not to compromise the authenticity or integrity of my role as researcher.

Figure 5.2. The maypole. In this school, playground development had a high profile. Here a teacher uses a much loved cherry tree as a maypole around which they dance to celebrate May 1st. The absence of grass underneath the tree indicates how well used the area was throughout the year for running around and playing underneath.

The inclusion of the photographs in this chapter is designed to expose the epistemological problem of person-place interaction. Changes to spaces are connected to a range of possible effects socio-culturally and in terms of learning. The collaborative approach to the co-construction of photographs is on-going; readers are invited to engage in their own analysis. My early analysis followed Collier's (2001) basic model. The photographic data was initially arranged in a catalogue according to location. The first stage involved a form of open viewing of all the visual data. After an inventory was made, a more structured analysis was done and, lastly, a return to the complete visual record was necessary in the search for significant meaning and triangulation with other findings.

During the stage of open viewing, one insight emerged from analysing the photographs. The quite barren environment of the playground, with few recognisable "play" installations, provided an unexpected number of opportunities for a wide variety of play activity. My ability to spot these was enhanced by my continued presence in the field, interaction with the children and the receipt of explanations for what was going on in their terms. A number of distinct areas in the playground became identifiable as central sites for different phases of play. These could emerge quickly and just as easily disappear as the playtime progressed. It seemed that these outwardly static hard surfaced tar-mac play areas continually changed over time *through use*, being shaped and reshaped anew with each subsequent break time, season and participation by different groups of children.

Sustained presence in the field confirmed that person-place interaction is a joined-up process related to how materials and forces (imagined and real) are ordered and reordered over time. In order to pay due regard to this assumption, activity was therefore coded as person-place interaction patterns. Patterns arose from interaction between artefacts that seemed quite insignificant. Places were not so much used as provided opportunities for, or even actively invited, certain activities once combined with the malleable rules for play, children's imaginations, and other elements of the network. Almost all of the spatial and material features that became central in play activities were designed with children in mind. Some examples included: signposts were used for swinging around. (One become known as "the dizzy pole"); corners invited hiding games or places for telling secrets; markings on the ground areas allowed smaller groups to gather in a "safe zone" or den during games; and elevated places afforded opportunities as viewing points.

The role natural objects played in these patterns drew my attention to the diurnal and seasonal effects. The patterns of activity were dependent not just on what the children decided they would do (human agency) but on the availability, pliability or potential for engagement these other material bits and pieces offered. Agency was distributed across the network of heterogeneous materials that went into the playground activity. Bushes, berries, sticks, stones, frost, leaves, ice, mud, and trees all played parts in children's engagement in play and social interactions of various kinds. For example, the coming of the blackberries provided objects for throwing and catching. Younger children used particular berries for marking their hands. These served as identifiable stamps that allowed children who were members to gain access to bush dens. The relational effects of popular culture (the trendiest of activities), the effects of seasons, weather, and even the agreed presence of imaginary others all played their part. There were also more obvious power effects: the sense that some activities were deemed permissible or that certain places were out-of-bounds and who in particular was in supervision. Teachers appeared to be the most significant agents in getting projects going. They also functioned as gatekeepers on participation by children and outsiders. These human and non-human effects combined to bring forth the place we called the playground and the social life that was found there. Places appeared and disappeared wherein identifications were made available or rendered impossible as opportunities for these children.

The second phase of the project involved at least one visit to each of 22 other schools. The criterion for selection was their reputation among practitioners for having involved children in decision-making and participating in change in school grounds. Nineteen of the schools were primaries, two were secondary schools and one was an urban nursery school. Twelve of the primary schools were small and in rural settings. There was one larger rural school. The schools came from six different regions in the central belt of Scotland and the Grampian-Highland region. During school visits I usually entered classrooms first where I adopted an outsider-facilitator role in order to conduct a survey of children's views of their perceived degree of involvement in decision making in projects. This involved using a ranking exercise similar to those used in forms of rapid rural appraisal and participatory rural appraisal (Chalmers, 1992). In most schools visited, a team of children were selected for their high levels of insider knowledge about various grounds development. They took me on guided walks of grounds while providing a commentary on the purposes and processes involved in their development and the manner in which children used these new places. I took 195 photographs while on these walks. Sometimes these shots were collaboratively constructed with children either voluntarily or spontaneously demonstrating their new favoured activities associated with the changes "on the ground."

Different school grounds projects sought different socio-spatial rearrangements (such as a pond, or a seating area), all of which emerged from a sharing of the problem, for example, the need for somewhere for wildlife or somewhere to sit and eat. The various rationales put forward by school staff is revealing. They included the need to deal with bullying, to address the aesthetic impact of the school site, to improve opportunities for play, to increase the diversity of wildlife present, to deal with issues of safety, and to increase the potential of the site as a teaching resource (Kenny, 1996). The diverse range of possible collaborators who would become engaged (such as the Prince's Trust, Scottish Natural Heritage, etc.) revealed other aspects of the motivation for change and the positioning of children, schooling and learning. Other motivations included hopes for improving school-community links and the advancement of parents as more direct partners in school life. The context for the analysis of these changes to school sites is that there is a perceived public, organizational or interpersonal need for it. Developments can, therefore, be seen as efforts to create spaces for alternative cultural practices, wherein identities and learning can be opened up. These rationales mask quite hopeful views about the 'place' of children in society and what is regarded as consequential learning and adult-child relations generally. These hopes were spatially expressed in the motivations for change and also in the designs and the physical reconfigurations made. The efforts to envision better and new places into being can be called "utopic practice" (Marin, cited in Hetherington, 1998). While nowhere can actualise Utopia itself, places can symbolise aspects of different utopic practices. The very activity of planning is an act of hope in itself.

The diverse approaches to the changing school grounds portray different essential fictions of the child-as-participant. School grounds, therefore, are examples of sites within educational discourse that display the elements of the different utopics. These identifications position children as different types of learners/actors/subjects. The analysis of photographs as re-presentations of alternative spatial practices permitted seven recurrent utopic patterns to emerge. The seven patterns were the utopics of the outdoor classroom, of safety, romantic, communitarian, of citizenship, of sustainability, and of the tribal child (see Table 5.1). These utopics offered different identification opportunities for the participant child, respectively the child in need of teaching, in need of protection, who should be happy, who symbolizes community hope, which is citizen, who is eco-activist and the tribal child. In practical terms, this might have meant opportunities for children to carry out experiments about wind, do some design work with parents, grow and sell some vegetables locally, compost their lunchtime waste or to be involved in painting a fence. It also involved the repression or exclusion of opportunities for others, for example, the need for children to leave letter writing and bids for funds to adults because it was deemed inappropriate for children. The presentation of this typology is not meant to establish a reified version of all school grounds developments. There were many rich subtle differences between sites, their purposes and expressed values. Neither does this typology expose how influential the different utopics were across the schools visited. In many cases, the utopics that were mobilized overlapped in any one school or project. As a result, the analysis presented in Table 5.1 is a very distilled and simplified one but still worthwhile as a strategically interpretative rhetorical tool.

Table 5.1. A Typology of Utopic Patterns

The Identifications Mobilised	*Utopics of Children's Participation*
The child-in-need-of-teaching	Utopics of the 'Outdoor Classroom': Play-spaces become colonised.
The child-in-need-of-protection (The child in need of control and direction)	Utopics of Safety: design a safe place to play and work; children must be under supervision. Participation is rule-bound; management and surveillance of danger.
The child-who-should-be-happy	Romantic Utopics: Adults keep the difficult, contested or political aspects away from children—it's not their concern
The child-who-symbolises-community-hope	Communitarian Utopics: Doing something for the children becomes the motivation for community development, school survival or celebration
The child-who-is-citizen (The 'minority group' child who can make a difference)	Utopics of Citizenship: Adults work to simulate or make 'real' opportunities for children to have 'a say'.
The eco-activist child	Utopics of Sustainability: Children work to address biodiversity and conservation issues.
The tribal-child	Utopics of the 'Tribal Child': Children need space away from adults to 'do their own thing'—unstructured free play and children's 'own cultures' are important

Outdoor classroom utopics carried many of the norms of indoor educational practice into the school grounds. Subject disciplines remain strong and learning is constructed via the foundational knowledge generated through science, geography, maths and so on. The utopics of safe futures involves emphasizing the need to protect children from harm and danger. As a result, children were precluded from learning through activities that were deemed too dangerous because of health and safety concerns or from the perceived risks of working with outsiders or even parents. Romantic utopics carried with them the concern for children's "right" to a happy or even traditional childhood. This approach ran the risk of smothering learning in the desire to protect children from adult concerns, which would perhaps intrude on their innocent and happy childhoods. Communitarian utopics encourages contact with non-school staff and parents. In one case, the development of school grounds was part of a ploy to arrest the plans for school closure. Children learned from engaging in a variety of child-adult relations. At times however, some high levels of community involvement could mean the occlusion of children's participation.

Figure 5.3. The digger. In small rural schools, such as this one, grounds development usually involved a range of school-based staff, children and local people in collaborative efforts to make changes. Children's agency would fluctuate throughout a project such as this depending on the perceived degree of dangers, the nature of the activity and the pragmatic need to get things accomplished.

Figure 5.4. Compost. This child points to the skin of his recently eaten banana peel that sits on top of the school compost heap. All biodegradable waste in this school was composted in this manner.

The utopics of citizenship and sustainability were usually combined. They afforded learning opportunities that were linked to action projects that were seen to make a difference to society or the environmental health of the planet. Projects emphasizing these aims tended to be also concerned with the need for children to have a say and be active agents of change in partnership with adults. In one school's grounds, children grew herbs and vegetables that were later sold locally at a market. Children from one project recollected that design work involved more than "just letting your imagination run wild" and understood that impediments like financial constraints can effect design solutions. One child commented that the project "made you feel like a professional cos it might happen." Another said, "you don't just stay in the same position, you have a new role." By contrast, the utopics of the tribal child emphasized the need for children to have their own spaces away from obvious adult supervision and control. The hope was that children's learning would evolve within child-child spatial relations away from those seeking to manage children's time and space (see Figure 5.1).

RECONFIGURED SPACES

Spatially, the traditional school "yard" has been designed like a barracks. Interpersonally, the generations and even year groups have been isolated by prevailing patterns. Child-adult relations centred on rule making, pupil control and boundary issues. The janitor and the Council have traditionally been there to manage and maintain the site. The traditional school grounds site reveals a pattern designed to optimise the *indoor* classroom work ethic through the release of energy *outdoors*. Through development, the tarmac is often broken up and additional areas are incorporated. The "prison exercise" model is displaced through the introduction of features such as loose play objects, wildlife areas and amphitheatres. In developed grounds, elements of the carnivalesque are less repressed or ignored. Rigid straight lines are replaced with curves, labyrinth-like shapes and winding route ways. In terms of participation by adults, a diverse range of outsiders often become involved in the spatial practice of change. Adults from the community come in and children work side-by-side with them in negotiating many of these changes. In contrast to the bleak play areas of traditional sites, in developed grounds; the social well being of the whole community is addressed. Holistic health and quality of life are seen as important. As such, it can be argued that grounds changes are part of a new social movement to address the perceived need for the greening of public space. Many project activists recognised the need for a form of post-hierarchical politics. There was evidence in the photographs that hierarchical control, low maintenance, low cost minimalist purposes were overturned in favour of the ethic of care, personal and ecological health, biodiversity, community participation, and diffused agency among adults and children.

Laclau (1990) argues that the reinterpretation and contestation of space is increasing and with this process come new myths about who we may be. The analysis of school grounds developments has given weight to this assertion. For schooling, the contest over space provides a different lens on the construction of curricula. We have seen how different utopic patterns offered diverse contexts for learning. It was the active and spatial participation in change processes that made learning significant. The curriculum, insofar as it can ever be realistically visible, was not constructed in a predetermined manner out of formal, text-based lists of content. Instead, it became known through practice in local places with diverse actors: people, reconfigured sites and the bits and pieces of our locale. The curriculum arose within places of engagement that enabled different collections of people, and many other heterogeneous entities (trees, compost and toys and so on) to co-specify each other. Opportunities for new forms of identification for children and adults were made available as new patterns emerged.

Perhaps the word curriculum is another noun that needs reappraisal. After Newman and Holzman (1997), to see curricula ontologically rather than epistemologically means a shift from knowledge to knowing. If it was the interactions between the material bits and pieces, the people and the utopic desires that catalysed the learning (and vice versa), then our notions of curriculum may need to take on a similar, slippery awareness. To do so will require that we conceive of curriculum as a verb or pattern of purposes, possible identifications, spatial arrangements and events. Such a spatial reading of on-going curricular development would bring a recognition of the relational, the social and political. It would take us beyond seeing curricula as the formal, text-based documentation of content to seeing it as a sequence of learning-events-in-places. If this is so, then we need to take on board the importance of environmental and material aspects of learning. For experiments of participatory identification to occur, it would appear there is a requirement to approach curriculum development in a more open-ended manner. This approach must take cognisance of the spatial and the material. Learners need places of engagement, materials and experiences with which to build an image of the world and of them; and ways of having an effect on the world and making their actions matter (Wenger, 1998).

In many projects, the learning mobilized was distinctly different from that on offer in traditional schooling. Critically, it did not appear that these changes were happening across the whole network of schools. However, in many cases visited there was a sense that the projects were effecting whole-school culture. It is interesting to note that it was the school's *outdoors* that appeared to take on significance as a site of "border crossing" by parents, children and educators. Citizenship, community and ecological activism were literally emerging as important at the margins of school sites and school culture. The cross-case analysis of the complete set of 387 photographs revealed displacements and mobilizations in terms of time, space, interpersonal relations, utopic spatial practice and values, all of which had consequences for learning. Before developments, many school grounds were largely designed as spaces for energy release. Playtime was temporally compartmentalized into periods of fresh air *between* learning in classrooms. With developments such as the outdoor classroom pattern, this binary is disrupted. The value of slowness instead of speed is also asserted: quiet areas are introduced, informal learning is encouraged and the incidental and seasonal are celebrated. In many of the cases I visited, adult, male, managerial values were often challenged spatially via the expression of child-friendly, familial and collective values.

As far as this text is concerned, I invite you as reader to evaluate it for reflexivity and performative validity. Interestingly, I, as author, am absent from the photographs I include but I hope I have not hidden from your view entirely. I am behind the camera but my view is not a view from nowhere; my position is entangled in the framing of the shot and the authoring of this chapter. Educational research can enable a catalytic, if ambiguous interpretation of learning while still accepting the unavailability of a fully shared truth. Bringing to the fore the performative aspects of textual production and research brings out the possibility of a consciousness of performance. Critical spectators are necessary for good performances.

REFERENCES

Anderson, P. (2000). Children as researchers: The effects of participation rights on research methodology." In P. Christensen & A. James (Eds.), *Research with children: Perspectives and practices.* London: Falmer Press.

Chalmers, R. (1992). *Rural appraisal: Rapid, relaxed and participatory.* Discussion Paper 311. London: Institute of Development Studies.

Collier, M. (2001). Approaches to analysis in visual anthropology. In T. van Leeuwen (Ed.), *Handbook of visual analysis.* London: Sage.

Davis, B., & Sumara, D. J. (1997). Cognition, complexity and teacher education. *Harvard Educational Review, 67*(1), 105–125.

Denzin, N. K. (1997). *Interpretive ethnography.* Thousand Oaks, CA: Sage.

Fenwick, T. (2001). Work knowing "on the fly": Enterprise cultures and co-emergent epistemology. *Studies in Continuing Education, 23*(2), 243–269.

Hall, S. (1996). "Introduction: Who needs 'identity'?" In P. duGay & S. Hall (Eds.), *Cultural identity.* Thousand Oaks: Sage.

Hetherington, K. (1998). *Expressions of identity: Space, performance, politics.* London: Sage Publications.

James, A., & Prout, A. (Eds.). (1990). *Constructing and reconstructing childhood: Contemporary issues in the sociological study of childhood.* London: Falmer Press.

Kenny, K. (1996). *Grounds for learning—a celebration of school site developments in Scotland.* UK: Southgate Publishers.

Law, J., & Urry, J. (2002). *Enacting the social.* Centre for Science Studies and Sociology Department, Lancaster University at: http://www.comp.lancs.ac.uk/sociology/soc099jlju.html [visited on 6th June, 2002]

Lincoln, Y. S. (1995). Emerging criteria for quality in qualitative and interpretive research. *Qualitative Inquiry, 1,* 275–289.

Mannion, G. (2002). Education as a spatial practice: The shift from nouns to verbs. Contribution to conference symposium: *The altered spaces of lifelong learning.* SCUTREA Conference, Stirling: University of Stirling.

Massey, D. (1994). Double articulation: A place in the world. In A. Bammer (Ed.), *Displacements.* Bloomington, IN: Indiana University Press.

Mouffe, C. (1998). Radical democracy: Modern or postmodern?. In A. Ross (Ed.), *Universal abandon?: The politics of postmodernism.* Minneapolis: University of Minneapolis Press.

Newman, F., & Holzman, L. (1997). *The end of knowing—a new developmental way of learning.* London: Routledge.

Schechner, R., & Appel, W. (Eds.). (1990). *By means of performance.* New York: Cambridge University Press.

Stake, R. E. (1995). *The art of case study research.* Thousand Oaks: Sage.

Thrift, N. (2000). Afterwords. *Environment and Planning D: Society and Space, 18,* 213–255.

Wagner, J. (1979). Photography and social science process. In J. Wagner (Ed.), *Images of information. Still photography in the social sciences.* Beverly Hills, CA: Sage.

Wenger, E. (1998). *Communities of practice: Learning, meaning and identity.* Cambridge: Cambridge University Press.

CHAPTER 6

THE HIDDEN GEOMETRY OF CURRICULUM

Brent Davis and Dennis Sumara

In a recent study of mathematics learning, we worked with several teachers of young adolescents to structure and interpret their students' classroom experiences. In the first phase of the project, completed in 2001, our collective worked together on a four-week fractions unit. Two weeks into that collaboration, it was clear to all involved that almost every learner's understanding of rational numbers concepts already surpassed the curriculum's clearly stated behavioral objectives. Despite this consensus, however, at the meeting to debrief the events of the second week, one teacher opened with the comment, "I'm amazed at what my students can do. But I'm a little worried that we're not going to cover all of the topics, so I think I might use the last two weeks to go through them in order."

The emergence of the issue of "coverage" was not unexpected, especially given the fact that a high stakes, state-administered standardized test was scheduled for the end of the school year. What caught us off guard in his remark was the concern with "order" in the movement through fractions concepts. A very deliberate (and very explicit) aspect of the work had been a challenge of the tendency to parse and linearize topics, opting instead to structure lessons around activities that required a range of concepts to be applied at the same time. As project leaders, we had assumed that the students' demonstrated understandings would obviate desires to fragment and sequence of concepts.

Space, Curriculum, and Learning, pages 79–91

The teacher's concern, however, served as an important reminder that we, along with every other citizen of Western culture, are immersed in linearized worlds—conceptual and physical. In other words, we make certain spatial assumptions about the world which impact upon the way in which we act. Through this chapter, we aim to elaborate this assertion with regard to matters of curriculum. We also aim to develop some preliminary thoughts on alternatives to line-based structures and habits of interpretation in formal education contexts, in effect, to explore how different geometrical understandings of spatiality inform curriculum practices.

RIGHT AND REGULAR

More specifically, we might describe our project here as an effort to elaborate the post-modernist assertion that Western practices of interpretation have tended to be too rigidly constrained by the demands of rational argument. Our present concern might be read as a complement to that well developed critique. Alongside the cultural embrace of the logical argument there has been a more tacit, but no less pervasive tendency to incorporate particular geometric forms into conceptual and physical spaces.

Descartes' original insistence on the supremacy of the logical argument was rooted in his reading of Euclid's *Elements*. That text, assembled two millennia prior, was in essence an effort to collect and systemize the geometry of ancient Greece. Euclid's advance was a demonstration that an already substantial body of knowledge could be reduced to the intertwining consequences of a few elementary principles. Descartes was not the first to be impressed by the power of the rational argument. In fact, as is evident in the works of Plato, it predates Euclid as a model of thought in natural philosophy. In addition, formal logic was key to the intellectual revival of the middle Ages, foreshadowing the conceptual revolution wrought by Descartes and his contemporaries.

Euclid's major contribution was thus not in the mode of argumentation, but in the coupling of that mode to a set of forms and images that have since been embodied in many aspects of Western activity. In brief, through Euclid, the influence of rationalism extends well beyond the conceptual into the physical. In fact, as Mlodinow (2001, p. 96) develops, until the 19th century, "every work of architecture, the composition in every drawing and every painting, every theory and every equation employed in science were all inherently Euclidean." In other words, an investigation of the pervasive presence of classical geometry should not be limited to the contents of texts in university libraries. The shelves, aisles, halls, buildings, and campuses that house those texts are as thoroughly rational as any formal argument.

The straight line, in particular, has a particular prominence in Western worlds, as we hope might be evident in (the very Euclidean) Table 6.1. In it we scan the roots and some of the branches of several English words associated with straight lines. The Table is hardly exhaustive. We *strictly limited* ourselves in this *project* to vocabulary that is *explicitly aligned* with *prominent* Euclidean notions. In *truth*, we could easily *justify* an *extended list* that includes many, many other *ordinary* terms whose ancient roots and contemporary associations have to do with lines and linearities, including each of the words italicized in this paragraph.

Our point here is not that the emergence of this particular web of association represents some sort of error or conspiracy. It is, rather, that these terms are pervasive. They are present in English-speakers' communications and infuse habits of interpretation. Thus situated, they do a particular sort of work in the projection and maintenance of a "right" and "correct" sense of how things are. Beneath the literal surface of these terms is a mesh of rightness and wrongness, or correctness and falsehood, of straightness and queerness. In English, straight lines are knitted into how we think about good and evil, truth and deception, morality and deviance. Consider, for example, the stigmas of a few alternatives to straight: crooked, twisted, and bent, kinky.

Table 6.1. Some Straight-line-based Notions

Modern term	*Derivation*	*Some current usages and associated terms*
right	> Latin *rectus*, straight	right angle, righteous, right-handed, right of way, right/wrong, human rights
rect-	> Latin *rectus*, straight	rectangle, correct, direct, rectify, rector, erect
regular	> Latin *regula*, wooden straightedge	regulation, regulate, irregular
rule	> Latin *regula*, wooden straightedge	ruler, rule of law, rule out, rule of thumb, broken rule
line	> Latin *linum*, flax thread	linear, time line, line of text, line of argument, linear relation, sight line, linear causality, toe the line
ortho-	> Greek, *orthos*, straight	orthodox/unorthodox, orthodontics, orthogonal, orthopedic
straight	> German *streccan*, stretch	straight up, go straight, straight answer, straight talk, straight and narrow, straight-laced

The modern school has been particularly underpinned by linear assumptions, as might be expected, given its status as both creation and instrument of Western culture. Over the past few centuries in particular, discussions of curriculum and curriculum development have tended to be framed by what might be described as a rectilinear sensibility. These projects have been developed, uncritically, around metaphors of lines, trajectories, rectangles, and grids. Such metaphors seem to operate in at least two ways. On the more explicit level, they are woven through a variety of organizational strategies. In matters of curriculum, their presence is perhaps most obvious in those projects that are aimed at direct, unambiguous, accumulative, and age-appropriate tours through neatly dissected collections of concepts. On this count, linear and rectangular forms work in more than the conceptual realm as a strategy for organizing programs of study. One might also point to their more important influence on the physical structures of modern schooling, perhaps most apparent in the rectangular schedules taped to the corners of rectangular desktops arranged in rectangular rows within rectangular classroom stretched out along rectangular hallways.

The interpretive priority of the line is embedded in narratives of causality, certainty, optimality, progress, and efficiency. These ideals are woven through many texts and discussions on contemporary schooling. Indeed, this issue was central to the critiques offered by "reconceptualist" curriculum theorists in the late 1960s and early 1970s (see Pinar, 1975), as they offered a response to the formal, rational, and very linear structures of curriculum development that dominated discussions of curriculum development in North America (Bobbitt, 1924; Tyler, 1969).

The hold of linear forms is not strictly a matter of a cultural preference for Euclidean forms. Pragmatically speaking, for example, a key reason that linear mathematics held sway in classical times was the fact that it lends itself to calculations that could be done with pencil and paper. Descartes, Newton, and their contemporaries were well aware of nonlinear phenomena. However, as Stewart (1989) suggests, because of the intractability of many nonlinear calculations, when they arose they were immediately replaced by linear approximations. This included such phenomena as growth and co-dependencies of living forms. Stewart (1989, p. 83) explains that this "habit became so ingrained that many equations were linearized while they were being set up, so that the science textbooks did not even include the full nonlinear versions. Consequently most scientists and engineers came to believe that virtually all-natural phenomena could be described by linear equations. As the world was a clockwork for the 18th century, it was a linear world for the 19th and most of the 20th century." Stewart also observes that the situation has recently changed, with the advent of more powerful calculation technologies. Across the sciences, the universe has come to be seen as "relentlessly nonlinear".

Notably, the still deeply entrenched scientific preference for line-based interpretations of phenomena appears to be more than a matter of efficient calculation. The power of linear relationships and line graphs may arise in the fact that the human mind easily recognizes certain shapes, lines in particular. There is considerable evidence that such tendencies are constitutional and not simply cultural. They have, for example, been linked to the disproportionate cognitive resources given over to vision within the human brain. Stewart and Cohen (1997, p. 187) assert, "the standard raw materials of geometry, the notion of a point and a line, match the physiology on our visual systems closely." Others have pointed to the evolutionary advantages that come with the predilection to simplify the world through causal explanations and efficient action schemes (Ehrlich, 2000). Still others have highlighted that, from the few hours of birth, humans are seeking out lines and sensory systems are exaggerating edges of objects (Hoffman, 1998). It does not seem unreasonable to expect that such tendencies would be carried into the articulation of more abstract competencies. As Cohen and Stewart (1994, p. 369) put it, "drawing fine lines is a human tendency, an attempt to make our simplified mental labeling system match a differently structured world." We return to the matter of the "differently structured world" shortly.

NORMAL AND STANDARD

The pervasiveness of lines and linearities comes into even more dramatic relief through a similar analysis of origins, meanings, and current associations of terms linked to another key Euclidean form: the *right* angle, and, in particular, notions that have come to be clustered around the term *normal* (see Table 6.2). *Norma*, the Lain root of normal, originally referred to the carpenter's square, the instrument used to measure and to create right angles for buildings. According to the *Oxford English Dictionary*, when the word was incorporated into the English language several centuries ago, *normal* was used to describe the accuracy of a cut or a corner. Something was normal if it was sufficiently close to 90 degrees for the purpose at hand. By the early 1800s, people had begun to use normal to refer to other measurements. A figure or form was described as normal if it matched with predetermined specifications of some sort. This idea was of mounting importance in an era of rapid industrialization and mass production. In particular, there was an increased demand for uniform (read: normal or standard) parts for machines, since these were much less costly than individually manufactured pieces.

Table 6.2. Some Right-angle-based Notions

Modern term	*Derivation*	*Some current usages and associated terms*
Standard	> Latin *stare,* stand (i.e., make a right angle to a planar surface)	standardized tests, standard form, raising standards, standard time, standard units, standard deviation, standard of living
Normal	> Latin *norma,* carpenter's square	normal curve, normalize, normative, normalize, normalcy, normal fork, normal child

This need for standardization supported the emergence of a new field of inquiry, statistics, which drew on developments in mathematics and several applied sciences, such as astronomy, geology, and biology. By the mid-1800s, a number of statistical methods had been developed to help cope with scientific and industrial needs. Once developed, those tools spread rapidly in academia. Researchers quickly realized that the distributions of many natural phenomena could be reasonably well described in terms of one or another distribution curves, most prominently, the "normal" or "bell curve." In other words, the idea of "normality," which was already based on a metaphor out of Euclidian geometry, became re-mathematized. In this re-iteration, *normal* became precisely measurable, defined in terms of nearness of a particular data point to a calculated mean value.

In the latter half of the 19th century, researchers in psychology, sociology, and related domains picked up on the normal curve as a possible route to scientific legitimacy. Tools that were rooted in notions of linearity and that were invented to describe variations within natural populations were employed in the creation of such phenomena as intelligence quotient, aptitude, and so on. By 1900, supported by these tendencies, the idea of a "normal man" (normal in terms of both development and attributes) was in wide circulation. (For more thorough discussions of this evolution and some of its social implications, see Davis et al., 2000; Foucault, 1990; Kline, 2001).

Supported by this new, normalizing "science" of measurement and development, education policy-makers instituted some sweeping changes to the structures of schooling in the first half of the 20th century. In a word, things became much more *standardized.* (Note the roots of the term in Table 6.2). Common curricula, homogeneous classrooms, uniform teaching methods, age-appropriate routines, and standardized examinations soon became commonplace. The line and right angle won out. Virtually every conscious aspect of learning, intelligence, and schooling was lined up, standardized, normalized, and/or graduated/graded.

NON-LINEARITY AND NON-NORMALITY

One of the problems that arises when such notions as *straight*, *right* and *normal* are coupled to their opposites (e.g., *bent*, *wrong* and *abnormal*) is that such pairings are often assumed to span the full range of interpretive possibility. This is a problem with line-based systems of thought. Once two poles are defined, all events or claims are presumed to fall somewhere between them. Dewey (1910, p. 19) described the situation in this way:

> Old ideas give way slowly; for they are more than abstract logical forms and categories. They are habits, predispositions, deeply engrained attitudes of aversion and preference. Moreover, the conviction persists—though history shows it to be a hallucination, that all the questions that the human mind has asked are questions that can be answered in terms of the alternatives that the questions themselves present. But in fact intellectual progress usually occurs through sheer abandonment of questions together with both of the alternatives they assume—an abandonment that results from their decreasing vitality and a change of urgent interest. We do not solve them, we get over them.

Consistent with Dewey's analysis, if one were to survey trends in scientific research over the past several decades, there has been a pronounced movement away from assumptions of linearity, particularly around matters of living and learning systems (Capra, 1996). With this shift from a line-based geometrical understanding of spatiality to understandings rooted in more convoluted or unruly forms, there has been a change in the status of mathematics. Mathematical descriptions of nature are no longer cast as fundamental truths, but as models or possibilities.

Prominent among these new models are images drawn from two intertwining branches of mathematical inquiry, namely systems theory and fractal geometry, both of which are subsumed within what Capra (1996, p. 35) calls "network thinking." He explains:

> the web of life consists of networks within networks. At each scale, under close scrutiny, the nodes of the network reveal themselves as smaller networks. We tend to arrange these systems, all nesting within larger systems, in a hierarchical scheme by placing the larger systems above the smaller ones in pyramid fashion. But this is a human projection. In nature there is no "above" or "below," and there are no hierarchies. There are only networks within other networks.

The images used by Capra represent a challenge not just to the images that have tended to remain tacit in modern descriptions of phenomena, but of the assumption of the power and supremacy of reductive logic. Whereas Euclidean forms can be reduced to fundamentals and basics, the geometry of network thinking is one in which complexity is not a function of scale, but of the conditions of observation.

Or, phrased in terms of an emergent branch of mathematics, nature seems to display a certain fractal structure. A fractal is a geometric form that has intricate structure on all levels of magnification. Whether one zooms in or zooms out, the bumpiness of detail remains constant, in contrast to Euclidean forms. Fractals, that is, do not reduce to lines and points. In part because of this quality, fractal geometry has been described as "far closer to the flexibility of life than the rigidity of Euclid" (Stewart, 1998, p. 23). Indeed, a major impetus in the emergence of fractal geometry was a desire to describe and analyze the complexity of the irregular shapes in the natural world.

Fractal images present interesting alternatives to the sorts of forms that have been used to interpret and to structure aspects of schooling. Consider, for example, conceptions of personal knowing and collective knowledge. Overwhelmingly, such phenomena are entangled in line-based assumptions of fundamentals and basics, of directions and progressions, of constructions and trajectories. They also tend to be framed in terms of a certain tension between the individual and the collective, in which schooling is usually imagined to serve as a mediator of sorts.

A fractal-based interpretation would offer a very different spatial frame. Individual knowing, for instance, is not seen in tension with but as nested within collective knowing. Both are evolving, emergent forms that are comprised of and nested in other, similarly dynamic and comparably complex forms. Along with the shift in image comes a shift in the assumed dynamic. The common progress-oriented images of ladders, paths, programs, and so on are supplanted by images that present the ever-present spray of possibilities, and always expanding "space of the adjacent possible" (Kauffman, cited in Stewart, 1998). Indeed, a new vocabulary is emerging to describe such dynamic events, as the assumption of "goals" or "targets" at the end of linear trajectories are replaced by discussions of "phase spaces," which look less like roads cut across featureless plains/planes and more like a map of the range of possible paths a stream of water might follow on its way down a rugged mountain slope.

At first hearing, such phrasings may not seem to be of much immediate use when held against the physical and conceptual linearities that have been used to give shape to modern schooling projects. However, we would highlight that these sorts of sensibilities have been embraced by a range of educational theorists who have examined several of the nested aspects of modern schooling. To mention a few, radical constructivist accounts of individual sense-making (von Glasersfeld, 1994), situated learning perspectives on group identifications (Lave & Wenger, 1991), social constructionist accounts of collective knowledge (Ernest, 1991), and cultural theoretic perspectives on the evolution of social institutions (Walkerdine, 1988) share both an assumption of complex evolutionary dynamics and a sense (often implicit) of the nestedness and complicity of emergent forms.

There have been some prominent efforts to highlight the fractal-like characters of activities at some of these levels. Neurologist Calvin (1996), for instance, has described the clustered and nested structures of the brain. Anthropologist Aiello and psychologist Dunbar (1993) have extended such observations to the levels of the social and the cultural. Ecologists, in fact, have taken the discussion to the level of the planetary (Abram, 1996; Lovelock, 1979). With particular reference to the phenomenon of collective knowledge, researchers have recently developed "maps" of specific fields by tracing out citations and references. The resulting images are decidedly fractal, as nodes of influence cluster into and around larger nodes, and so on (Chen & Paul, 2001). The result is consistent with Capra's (1996, p. 39) description of a shift in conceptions of knowledge: "in the new systems thinking, the metaphor of knowledge as a building is being replaced by that of a network. As we perceive reality as a network of relationships, our descriptions, too, form an interconnected network of concepts and models in which there are no foundations." This conception of networks also serves to highlight an important quality of many dynamic events. For some phenomena, there are no "normal" cases, that is, statistics-based normal distributions are not applicable. The list includes earthquakes, avalanches, wars, fads, city sizes, and personal influence on collective activity (Buchanan, 2000). These and many similar dynamic phenomena seem to obey a "power law distribution," by which an event of double the intensity (or size or influence) is roughly four times less likely, triple the intensity nine times less likely, and so on.

One reading of the power-law is that there are no normal experiences. What's more is that the significance of an event cannot be predetermined, but only emerges after a sometimes extended period as it is amplified or muffled by other happenings. Framed in terms of the creation of a fractal image, a recursively elaborative process must be allowed to run its course before that image can be known. More colloquially, there's no way to distinguish between a future mountain and a molehill when a bump first appears.

Used as an analogy for the educational process, this cluster of ideas might be taken to suggest that the significance of any particular event of curriculum can never be fully appreciated in the moment of engagement. The same is true of all complex, emergent events. This is a realization, as

Capra (1996) points out, that has been "disturbing to scientists" educated in line-based traditions and that has prompted a shift in emphasis from quantitative to qualitative analysis.

THE NOT-NORMAL SCHOOL

Our intention in this writing is not to suggest that Euclid was wrong. On the contrary, we regard him as a luminary, a person whose work opened up a window through which otherwise unseeable aspects of the universe are revealed. The error has not been the development or the embrace of Euclidean geometry, but the taking literally (Rorty, 1989) of figurative notions rooted in Euclid. On this count, it is curious to us that so much attention has been focused in recent decades on the privilege of the logical argument, and so little to the implicit geometric structures that are used to give shape to that mode of argumentation. We suspect that a key aspect in the continued priority of rational thought is that it has become naturalized. More poetically, Abram (1996, p. 64) makes the point this way: "the super straight lines and right angles of our office architecture ... make our animal senses wither as they support the abstract intellect; the wild, earth-born nature of the materials—the woods, clays, metals, and stone that went into the building—are readily forgotten behind the abstract and calculable form." The challenge for educators is thus not a matter of exorcising linearities; it is about expanding possibilities through different senses of spatiality. On this count, it bears mention that fractal geometry is not a rejection of Euclid's geometry of the plane, nor for that matter of Descartes analytic geometry. It is utterly reliant on those branches of mathematics; it is a recursive elaboration, a thread amid a tangle of strands, not a correction or the root of a new worldview.

In our own work, we have embraced fractal notions because they serve the twofold function of helping to uncover some now-transparent figurative forms and for their usefulness in framing aspects of learning and schooling. Such modern curricular obsessions as orderly progression, efficiency, standardization, uniformity, and age-appropriateness, as embodied in such obsessions as spiral curricula and linear lesson plans, are rendered suspect in the realization of their associations with reductive images and logics.

In particular, we have found fractal geometric notions very useful to interrupt one of the most prominent topics of discussion among educators, at least in North America: the relative merits of teacher- versus learner-centered classrooms. We observe at least two troublesome Euclideanisms that are implicit in the terms of these persistent debates. First, "center" is one more notion that is borrowed directly from mathematics. A center is the midpoint of a well-defined geometrical object, a construct that does not seem well suited to so amorphous a form as a classroom collective. By contrast, a fractal has no center. Rather, fractals are better described in terms of nodes that are clustered into nodes (and so on), and we believe this image is much better suited to the multiple, dispersed sites of teaching-and-learning that are present in a classroom collective.

A second, subtler problem implicit in teacher- versus learner-centered debates is that both sides in these arguments rely on the reductionist assumption that the individual is the fundamental particle of cognition. Once again a fractal is useful as a visual metaphor for an alternative interpretation. Just as nodes cluster into larger nodes in a fractal form, so cognizing agents might be seen to cluster into larger cognizing agents. More bluntly, a social collective such as a classroom might be considered a learning system, a conception that presents some tremendous interpretive advantages to teachers and researchers. For example, one can actually observe the cogitations of the collective and, hence, can make pedagogical decisions that are informed by sound knowledge of emergent patterns and norms. As well, this manner of thinking is useful to make new sense of the teacher's role. Framed in fractal terms, the teacher is clearly a co-participant in the collective mind, one that is implicated in every aspect of learners' knowing. (A detailed discussion of an extended teaching experiment in which the classroom collective was explicitly framed in fractal terms is presented in Davis and Simmt in press.)

Such matters are frequently re-emphasized to us in our ongoing work with practicing teachers. For instance, after a relatively brief introduction to these sorts of images and ideas, a veteran mathematics teacher, Gordon Booth, commented:

> "Both we as teachers and our students view learning as a structured activity that happens in school where the components of what is to be learned is broken down into bite-sized pieces and delivered in a logical sequence so that the learner can reproduce the desired information to an acceptable standard. While this may look like a viable structure, any teacher will tell you that the path outlined in a curriculum guide is not the path that the teacher and the learners walk down. The interests and backgrounds that the students bring to the classroom, along with the activities presented by the teacher, help to direct the meandering path taken towards the prescribed goals."

When we first read Gordon's statement, our first temptation was to interpret it as an indication of a shift in thinking. Upon review, however, it is actually a case of using a new vocabulary to express some aspects of what experienced teachers already know, for example, that learning doesn't follow straight lines and that one's pedagogical attentions are better focused on the evolving classroom collective than on the individual learner.

Despite their conceptual significance, such elaborations do not seem to prompt radical transformations to classroom practice, in large part because such ideas are already embodied in the actions of experienced and effective teachers. However, where the ideas have proven particularly useful is in the redescription of curriculum. When framed in terms of a co-emergent, nested phenomenon, that is, effectively, a collective cogitation, curriculum becomes something far more than a walk along an established path. Curriculum itself becomes a dynamic, living form.

It is thus that we find ourselves following Rorty's (1989, p. 78) recommendation that such projects be considered in terms of redescription rather than logical argument. We imagine ourselves to participate in the invention of new vocabulary and the re-appropriation of established terms in the hope that "people will no longer ask questions phrased in the old words." Our purpose in the introduction of fractal images within a discussion of curriculum, then, is not to lay new vocabularies and images atop old ones. Our intentions, rather, are to provide means to raise issues and to pose different questions, ones that cannot be framed in classical terms. Following Dewey, we hope to provide means to *get over* certain worries, rather than to *solve* or *resolve* them. In this regard, we are hopeful that education, as a domain of inquiry, will soon join with economics, ecology, cultural studies, the medical sciences, and many other fields (Capra, 1996; Johnson, 2001) in the embrace of nonlinear forms to characterize their complex, situated, and emergent objects of inquiry.

REFERENCES

Abram, D. (1996). *The spell of the sensuous: Perception and language in a more-than-human world.* New York: Pantheon.

Aiello, L. C., & Dunbar, R. (1993). Neocortex size, group size, and the evolution of language. *Current Anthropology, 34,* 184–193.

Bobbitt, F. (1924). *How to make a curriculum.* Boston: Houghton Mifflin.

Buchanan, M. (2000). *Ubiquity: The science of history … or why the world is simpler than we think.* London: Phoenix.

Calvin, W. (1996). *How brains think: Evolving intelligence, then and now.* New York: Basic Books.

Capra, F. (1996). *The web of life: A new scientific understanding of living systems.* New York: Doubleday.

Chen, C., & Paul, R. (2001). Visualizing a knowledge domain's intellectual structure. *IEEE,* March, 65–71.

Cohen, J., & Stewart, I. (1994). *The collapse of chaos: Discovering simplicity in a complex world.* New York: Viking.

Davis, B., & Simmt, E. (in press). Understanding learning systems: Mathematics education and complexity science. *Journal for Research in Mathematics Education.*

Davis, B., Sumara, D., & Luce-Kapler, R. (2000). *Engaging minds: Learning and teaching in a complex world.* Mahwah, NJ: Lawrence Erlbaum.

Dewey, J. (1910). *John Dewey's "The influence of darwin on philosophy" and other essays.* New York: Henry Holt and Company.

Ehrlich. P. R. (2000). *Human natures: Genes, cultures, and the human prospect.* New York: Penguin.

Foucault, M. (1990). *The history of sexuality: An introduction.* New York: Vintage

Hoffman, D. D. (1998). *Visual intelligence: How we create what we see.* New York: W.W. Norton.

Johnson, S. (2001) *Emergence: The connected lives of ants, brains, cities, and software.* New York: Scribner.

Kline, W. (2001). *Building a better race: Gender, sexuality, and eugenics from the turn of the century to the baby boom.* Berkeley, CA: University of California Press.

Lovelock, J. (1979). *Gaia, a new look at life on Earth.* New York: Oxford University Press.

Mlodinow, L. (2001). *Euclid's window: The sorty of geometry from parallel lines to hyperspace.* New York: The Free Press.

Pinar, W. (Ed.) (1975). *Curriculum theorizing: The reconceptualists.* Berkeley, CA: McCutchan.

Rorty, R. (1989). *Contingency, irony, solidarity.* New York: Cambridge University Press.

Stewart, I. (1989). *Does God play dice?* Cambridge, MA: Blackwell.

Stewart, I. 1(998). *Life's other secret: The new mathematics of the living world.* New York: John Wiley and Sons.

Stewart, I., & Cohen, J. (1997). *Figments of reality: The evolution of the curious mind.* Cambridge, UK: Cambridge University Press.

Tyler, R. W. (1969). *Basic principles of curriculum and instruction.* Chicago: University of Chicago Press.

Walkerdine, V. (1988). *The mastery of reason: Cognitive development and the production of rationality.* London: Routledge.

CHAPTER 7

UNDERGRADUATE CURRICULA AS NETWORKS AND TRAJECTORIES

Jan Nespor

Current views of higher education, inspired in part by the spread of computer mediated forms of instruction, treat curriculum as synonymous with "content" and define teaching as the generation of content landscapes across which flexible students can pursue individualized interests. The first quote below is from the President of Teachers College, Columbia University, the second from a well-known scholar and theorist:

> Colleges and universities are in the business of discovering and disseminating content—information and knowledge—and today content is king. Digital technology gives television, telephones, and cable stations the capacity to distribute more and more content, and today there are more channels available to distribute content than there is content to fill them (Levine, 2001).

> The 200-year-old structure of the curriculum has finally begun to change.... Instead of ddivisions and departments comprising courses hierarchically ordered and sequentially arranged, imagine multiple nodes organized nonhierarchical with manifold connections to other nodes is a curriculum, which is constantly changing. Each node is formed by the intersection of many lines of investigation (Taylor, 2001).

Space, Curriculum, and Learning, pages 93–108

Claims like these conflate the uses and meanings of information artifacts with their material forms, and ignore the spatial and temporal contexts of knowledge-making activities in relations of power. I return to them at the end of this chapter, but invoke them here because they focus attention on the previously neglected topic—the spatial distribution of undergraduate curriculum—which serves as the focus of this piece. Curricula, I will suggest, are best thought of as means for connecting school-based events to events and situations outside the school. To flesh out this idea I draw on data from interviews conducted over two years at a U.S. university, Virginia Tech, with 60-some students enrolled in an introductory nutrition course. Most of these students were interviewed twice over a semester (early in a term, then at the end); some were asked to keep journals for a week, others time-logs, and these were used to structure interviews, as were photocopies of the students' class-notes.[1] The account of curriculum I draw from this material not surprisingly reflects a student "standpoint" (Smith, 1987). However, instead of privileging this view, my aim is to use it strategically to show the inadequacies of treating curriculum as static and despatialized, and to suggest some alternative concepts, which better register how it is spatially and temporally constituted.

EDUCATIONAL EVENTS STRETCH OUT IN SPACE AND TIME

Instructional events need to be characterized in relational terms: as assemblages (Cooper, 1999), or intersections (Rosaldo, 1989), their meanings "intertextually" constituted (Lemke, 1995); or to borrow from the geographer Massey's (1993) conception of place, as "articulated moments in networks of social relations and understandings" rather than "areas with boundaries around." As Hanks (1996) points out, "the presumed locality of the indexical present is extremely variable in actual interaction, and cannot be defined without reference to the sociocultural background of the interactants and the space in which their interaction takes place. Much of what defines the here-and-now is neither local in the operative sense, nor is it accessible to audiovisual recording or transcription."

Imagine, for example, a university student working in a dormitory room on a physics problem (it could be a problem in chemistry, accounting, mathematics, or many other fields). A number of space- and meaning-making connections intersect in the event. The space of the classroom is extended into the student's residence, and the student's out-of-school time is synchronized not just with the professor's pacing of course material, but also to an institutional calendar, which organizes "learning" into arbitrary units of time like semesters, and semesters into multi-year programs of study. There are also links between the space-time of the problem-solving and anticipated evaluative events elsewhere in the future. At the same time, insofar as the problem solving is part of a "course," there are connections to previous physics or math courses the student has taken where similar problem forms, tasks, and concepts were encountered. There are also connections to disciplinary sites where "physics problems" are constructed and warranted (networks connecting physics departments and publishing houses), and to physics courses around the world where the same or isomorphic problems are assigned. All these articulations are moving. To borrow a phrase from Hanks (1996), to be schooled "is to enter a field in which everything is moving and objects exist in relation to one another." The physics problem is not so much an "articulated moment" as a moving articulation, the translation of a succession of place-makings, enrollments, decontextualizations and recontextualizations, temporally unfolding from secondary school onward.

The point is not that educational events are stretched out indefinitely or to some point of incoherence—things aren't that frothy (Smith, 2000). However, whatever stability and boundedness obtains is a contingent accomplishment, an effect of calibrating and synchronizing events with larger-scale and relatively more stable networks or "actor networks" (Callon, 1986; Latour, 1987; Law, 1994; Law & Hasard, 1999). In this view, the meanings of an event are constituted by hooking it up to moving networks of people acting with, through, and by virtue of their entanglements with durable artifacts, structures, and materials. Into these networks of action are woven so many commitments, identities, and interests (in the form of material structures, equipment, legal rules, lines of funding, circuits of communication, organizational roles, and so on) that dismantling them, while always possible in principle, is in practice very difficult (Callon, 1991).

In undergraduate education there are at a minimum two (and usually many more) such calibrating actor networks: the academic disciplines and the university itself. Notwithstanding their differences, disciplines like physics, English, and the other subject matters, are networks of power-knowledge that define and regulate practice in domains of activity, including (though again in very different ways) the lines of educational preparation that aspirants must follow to become disciplinary practitioners. The university, meanwhile, generates a spatial and temporal order of its own as an organization managing a professional labor force, pressured by external "markets" and organizational fields, and accountable both for educating students and for developing research knowledge through the work of faculty and graduate students. Students move in both of these spatial orders, organizing themselves with reference to the flows of various disciplines, and constructing cumulative organizational identities. In addition, students must situate their academic work in various familial, workplace, and interpersonal networks. As I suggest in the next section, the meaning of any instructional event or artifact is constituted by the way it is contextualized in such networks.

NETWORKS AS CONTEXTUALIZERS

The nutrition course that served as the source of the student interviewees was a "survey" of the field enrolling 400–500 students each term. Enrollments per full-time faculty member are one of the university's measures of productivity, so from a departmental standpoint such large courses are devices for generating good statistics to buttress claims on university resources. Big courses are common at large U.S. Universities. Virginia Tech offers "as many as 50 classes each year with at least 300 students, 140 classes of 150 to 300 students and 170 classes of 100–150 students" (Zack, 1998). The larger courses, like Nutrition, have in recent years become targets for investments in computer-mediated instruction (Twigg, 1996). The professor teaching the course, in fact, received two grants from the university to work with instructional designers to create, among other things, Powerpoint slides that were posted on the course web page and doubled as class notes for students (e.g., Figure 7.1).

Such slides are not "school knowledge," curriculum, or "content." They are best thought of simply as objects that can be moved across space and time, as curriculum artifacts rather than curriculum per se. The slide belongs to a class of transportable objects that "render stretches of discourse discontinuous with their discursive surround" (Bauman & Briggs, 1990), as there are no linguistic or graphic devices, linking the slide to particular authors or historically- or geographically-specific locations (Silverstein & Urban, 1996; Collins, 1996; Olson, 1977). The terms on the slide are presented as directly and unvaryingly linked to universally available objects, processes, or states. It is school knowledge as timeless and spaceless truth.

And yet, as you look at the slide in the context of this chapter, these characterizations should appear problematic. Here, the slide is not about your body, or the diet you're planning for next week, or an item you have to know for tomorrow's test. It is an illustration of a certain discourse style in a chapter on undergraduate curricula. The meaning of the curriculum artifact inescapably changes as its context changes. The meanings of course materials are not timeless, spaceless attributes of the artifacts, but contingent outcomes of action; in Callon and Law's (1997) terms, "relational effects": "their form, their content, and their properties are not fixed. Rather their identity emerges—and changes—in the course of interaction.... Objects—for instance people and texts—are processes of transformation, compromise or

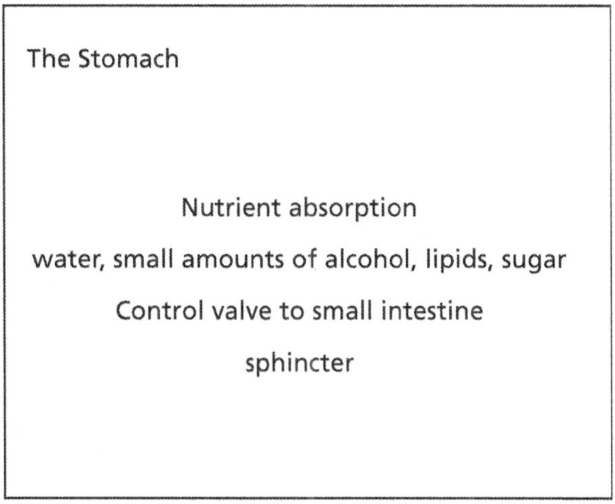

Figure 7.1. Slide from PowerPoint presentation

negotiation." Put another way, the slide doesn't just circulate in an abstract curricular space. It is always part of a course, or something to study for a test, or something to coordinate into a lecture, and so on. A thing, be it an utterance or an institution, can only be made meaningful, can only be recognized as "something" in the first place, by being situated in a particular context, a nexus of articulations that relate it spatially and temporally to other objects, events, and experiences. Making meaning, in Lemke's (1995) words: "can be represented as *selective contextualization*; we make sense of a word or deed by construing it in relation to other actual or possible words and deeds that we construct as the relevant contexts for interpretation."

Nothing remains "decontextualized," then, or as Bauman and Briggs (1990) put it, "decontextualization from one social context involves recontextualization in another." To make meaning (of the slide or anything else) we *necessarily* (no matter how much the work is hidden in our habitual ways of speaking and thinking) have to link it to specific spaces and times that we (and we assume others) can take for granted, *and* simultaneously connect it to events or ideas in ways that *produce* spaces and times. Meaning is thus a matrix of spatio-temporal relationships, some of which are assumed, others discovered, and still others negotiated or imposed. And it is not only objects like the Powerpoint slide that move through chains of spatial and temporal recontextualization. People also move, shift their spatial and temporal organization by attaching and reattaching themselves to more durable or extensive actor networks. For lack of a better term, I'll describe these on-going spatial and temporal realizations of context as "trajectories."

CURRICULUM AS TRAJECTORY

Schooling works by moving people and things along trajectories that ultimately situate them in spatial and temporal orders where only certain meanings, identities, and lines of action can be easily sustained. These trajectories are what I call "curricula." Curricula are not reducible to courses, academic fields, or subject matters, but are realizations of these things in the times and spaces of student and faculty practice. It follows that the multiple curricular trajectories threaded through universities each define a distinct spatial and temporal order, and in many cases begin (e.g., in secondary school) and end (e.g., in graduate school) outside the undergraduate years.

Like most transportation systems, curricula organize educational landscapes with maps and timetables that define some spots on the itinerary as places to stay and explore, others as detours or rest stops. Each program of study provides a list of the courses students are required to take, along with a list from which additional "elective" courses must be chosen. The lists include sample schedules that synchronize students' temporalities with the university's time (Munn, 1992), laying out a four-year program of study, term by term, that specifies all the courses students should take and when they should take them. These lists, and the longer students delay aligning their academic work with a list, the greater their risk of taking courses that will not count to the value of any particular course:

> "On the [course list] paper it has like all the classes you take your freshman, sophomore, junior, and senior year, and like for your freshman year it's like, it's tons of classes. Like that have to do with the major.... I mean it would have been nice if I knew last year ... I took Sociology and Communications. And for my major I have to take Psychology, so I basically took those two classes for nothing, and I have to go and take Psychology.... It is like a waste of time for me to take Sociology and Communications when I don't need them in my major, which it would have been better if I had taken other classes because it would have been easier to graduate on time." (S07, Food and Nutrition, 2nd year)

As another student explained, some degree trajectories leave students few options by requiring early commitments to narrow and tightly structured paths: "I know a lot of people who came in like Engineering their freshman year, and when you go Engineering your freshman year, they say, 'this is the classes you are going to take for the rest of your life', and so you really don't have any leeway to take what you want to do" (S15, Hotel and Tourism Management, 2nd year).

Since students often have little choice among courses required for the specialization field, they must use whatever time is left for "electives" to satisfy their curiosity, keep their grade point averages high—all while maximizing time for work in the major field of study. Students sometimes choose electives by seeking advice from more senior students who have already taken the courses in question, or, particularly in the case of students approaching graduation, on the basis of a course's perceived ease or difficulty. For example, S01, a finance student in her final year at university, took Nutrition because it met a distribution requirement for her major field, fit conveniently into her schedule, and, as an introductory course, was presumably easy and would help boost her sagging grade point average:

> "I took it because it was outside the Business school, which 50% of my credits have to be and because it fit into my schedule at exactly 1:00 which was when I had time to take a class. And I went through [the course timetable] and I said what BS [bullshit] class can I take at 1:00 pm that's a 1000- or 2000- level class [i.e., geared to 1st or 2nd year students and thus presumably easy] that I don't have to spend much time on?" (S01)

A given instructional event is thus not exclusively situated in a single curriculum: different curricular pathways can intersect in the "same" course or instructional event, splintering its meanings. The nutrition course, for example, was simultaneously part of a "pre-med" sequence for students trying to get into medical school, part of the training curriculum for hotel and restaurant managers, a throw-away elective taken for an easy grade by a business major like S01, and so on. Such courses thus fit into the academic careers of students in different ways depending on the curricular trajectories with which the students align. In any given semester these trajectories situate the course in different constellations of academic work. Consider the position of nutrition in the course-taking trajectories of S01 and S12, biology major, both in their last year as undergraduates:

S01	S12
Commercial Bank Management	Ecology
Intermediate Finance	Geology
Women and Creativity	Virology
Computing For Business	Microbiology
Nutrition	Nutrition

The schedules of the two students generate different daily, weekly, and semester rhythms, depending on the type and pace of tests, labs, homework, or project assignments. They organize students differently across the space of the campus (and non-campus dwellings), bring them into contact with other students of different ambitions and backgrounds, and entail the use of different means of representing and mobilizing course material for study. Like a word or phrase embedded in different utterances or texts, the nutrition course has different meanings and assumes different spatial and temporal forms in different curricular trajectories. For some students it is a co-present lecture course, for others an on-line course attended only on test-days: "my roommate she is in the same class as me, she never goes to class and her grades are better than mine.... Same class ... same time 9:00, and she skips it. She hasn't gone, the last time she went to a class was so long ago, like three weeks ago or something. But just knowing that the notes are on the Web she just gets them off of there" (S23).

Among students who actually go to the class, those interested only in the grade see the professor as an animator of the Powerpoint slides, adding nothing of relevance to the test. S08, for example, went to about two of three sessions, and claimed that the professor was

> "reading directly from her overheads and I can read, so I could print it off the Internet if I wanted to... She's reading verbatim like it's not even like she is really lecturing. She's just like, you know, the slide will show up and she will read it to us and we are like, OK, you know, we can all read, thanks. But I think it makes it a bit tedious and boring. Like if she would I guess customize it better to like just a lecture, to not sound like monotonous and like straightforward I guess."

By contrast, for students with some interest or background in biology or nutrition, S12 for example, the professor "doesn't just like go through the information and that's all she does. Like she stops and she asks questions and she just incorporates like, the class, like, to talk with her and stuff. And it just makes it a more laid back kind of setting so it makes it more fun for me I guess."

Curriculum artifacts like the slide are thus linchpins anchoring an instructional event in multiple space-times. The struggles over which space-time contextualizations are to "count" (or be registered in the organization's official account of its work) are at the heart of schooling. To teach is in part to connect instruction to particular events, situations, and streams outside the space-time of the official instructional frame (course, module, or lab). In this sense, "teaching" is not something that teachers alone do, but is an activity distributed across networks of people and things that extend beyond the university or school.

These networks differ in structure. Nutrition, for example, can be thought of as a region of the discipline of biology, and it can draw upon stabilized concepts from biology like those referenced on the slide—we don't have to cut someone open to demonstrate that they have a "stomach" that can digest "lipids." However, as a course designed to serve as an elective, nutrition must be structured to attract a large number of students from across the university.[2] The vast majority of students in the course (roughly 85%) are *not* specializing in fields like nutrition or dietetics, and would likely complain and stop enrolling if the professor focused only on the biology and chemistry of nutrition. Thus instead of moving students into a disciplinary space-time (for example, by having them work on discipline-generated phenomena like "lipids" in disciplinary environments like laboratories), the course is therefore designed to calibrate nutrition to students' current activities outside academic space-times. The goal of the course, according to one of the instructional designers working with the professor, was to see to it "that when the students finish the class they will be able to plan a nutritious diet for themselves."

The nutrition course presents an example of a course that can be fractured and recontextualized in terms of a wide range of trajectories. It is a different course for different students. This multiplicity is not inevitable, of course, but to constitute instructional events as more or less the "same" events for groups of students requires spatial and temporal organization on a scale far beyond the course. Curricula like physics or food microbiology, for example, can powerfully shape and standardize student contextualizations because they are densely embedded in tightly-organized, exclusive, world-wide disciplinary actor networks that exert considerable (though by no means complete) control over the objects and problems upon which students work. Students trying to become part of these networks are usually filtered through long sequences of tightly structured courses, segregated into physical spaces linked to disciplinary work spaces (e.g., labs), and configured into small, cloistered groups in which they learn to communicate and represent the world using the specialized categories and methods of the field (Nespor, 1994). Physics as a discipline extends its spatial and temporal organization into the space and time of the university, while in the case of the nutrition course, the university channels the space and time of the discipline into its organizational contour.

THE BOUNDARIES OF THE UNIVERSITY

In the case of programs like physics, curricula that organize students spatially and temporally with reference to networks "outside" the university articulate "local" practices like reading a book, working an equation, dissecting a weasel, or studying for an exam, with "globally" organized identities like "physicist," "veterinarian" or "accountant." Students thus don't have to construct from scratch ways of gaining access to such identities, jobs, and statuses. A biology student in her final year of a "Food Microbiology" specialization can make sense of her schoolwork as a tight, linear path to a specific professional future: "I want to go to graduate school and I want to, after that I want to like work in a lab or like work for some kind of food microbiology company. Like I could work for any kind of food company. Like Hershey's [a chocolate maker] or stuff like that. That's what I want to do" (S12). For such a student certain courses are *in* the space-time of the university as semester-long sets of requirements leading to a grade, but they are meaningful only in spatial and temporal frames that situate them *outside* the university. As this student explained, the relevance of a course was partly defined by what extra-university actors would make of it: "certain types of classes that you have taken they like to see. Like for me like it's important that I take like Microbiology, Food Microbiology kind of classes and stuff like that" (S12). In cases like this, a curriculum trajectory leading to a specific post-graduate identity largely overlaps with an official mapping of courses and other academic experiences. The curricular trajectories function as "infrastructures" (Bowker & Star 1999; Star & Ruhleder, 1996), technologies sunken from everyday view (like sewer systems and telephone lines, invisible to most until they break down).

By contrast, for the finance major quoted earlier, the connection of the list-defined program of study to a future identity was less certain. When I asked her what one did with a finance degree (she was less than a year from graduation at the time) she replied: "you know I have been asking myself that same question as have my parents and probably everyone who is graduating this year. No clue. Right now, what I am going to do? I guess you can go into stock broking or banking or doing financial analysis or becoming a financial planner or taking care of everyone else's money for the most part" (S01). Instead of a trajectory linking specific coursework and experiences to post-graduation identities, for such students the university mediates an indirect connection between coursework and post-graduation opportunities. Courses are relevant to students like S01 primarily insofar as they are translated into the grades or marks, which are then combined and translated into grade averages, and which the university then uses to regulate the distribution of diplomas. Diplomas function as a kind of "scale-jumping" (Smith, 1993) device organizing access to work and educational settings beyond the undergraduate college. Thus, the importance of instructional events within a given course varies over the semester as students figure out what they need to do to get the grades they consider acceptable, and the importance of the grade to be awarded for the course itself depends on when it is encountered along the curricular trajectory. As graduation approaches, grade point averages become increasingly salient and students judge courses according to how they are likely to affect this average. Access to job interviews, among other things, was regulated by grades: "everything is by your [GPA] right now, as I am finding out unfortunately" (S01).

BOUNDARY DISLOCATIONS

Massey (1996) has analyzed the "dislocation of the boundary between work and home" for men in high technology fields, in particular the invasion of the latter by the former. Such dislocations begin earlier for students in particular regions of university space-time. The term "extra-curricular" is often misleading when applied to students' jobs, personal relationships, activities on sports teams, or membership in fraternities and sororities. These are alternative contextualizing streams, actor-networks in their own right, coordinated with or existing in tension with academic curricular trajectories. In U.S. universities, for example, sororities and fraternities, in addition to serving as settings for degradation rituals, illegal drinking, and sexual depravity (Sanday, 1990), also act as appendages of the university's academic agenda, media through which old notes and tests circulate, and structures for disciplining students' academic effort:

> "I am our scholarship chair for our sorority which means I am in charge of getting everybody motivated to study and do all their tests and we do a little flower garden thing so if you get an A or a B you get a flower petal.... There are girls that have really bad grades, like below our average or whatever, and we are trying to set up a special program with them to get them more motivated, so if they can get their grades up so our whole chapter average will go up. So it's mandatory that they attend an hour and a half of the three nights that we have study hours." (S28)

In this case, life outside the official coursework is subsumed within curricular trajectories. Alternatively, students can align curriculum with their out-of-school identities or activities. For example, the choice of which curriculum trajectory to participate in is often coordinated with family- and peer-based networks. Many students inherit their major field of study. "I knew I was going to go into Business when I got here and I was leaning towards Accounting, but I didn't choose it until after I had my first couple of classes. My mom is a CPA and my brother followed my dad and became an Engineer, so I decided to go towards the mom side" (S5).

In other cases, students move over time from subordinating the demands of coursework to their non-school activities, to situations where their personal relationships are coordinated through curricular trajectories:

> "Fall semester I didn't do very good, and knowing that my boyfriend was coming here spring semester my parents like freaked out, and they were like, you are never going to be studying, and da, da, da. So I always felt like, especially spring semester, that I wasn't doing it for me, that I was doing it for other people, you know. And my boyfriend is a real good studier, and he

> really makes himself do it and last year, I didn't really care about it very much so he would harp on me too, you know, about you are not studying enough."

We come back here to the problem that the meanings of things change as their contextualization changes. Home, boyfriends and girlfriends, membership in sororities and athletic teams, living in a dormitory or a private apartment. All these things contextualize and shape the meaning of official academic work; as do the conditions of the larger public and private spaces in which students live and work as they study; the physical organization of the campus and area in which the campus exists, the availability of student-oriented or controlled spaces, the texture and accessibility of public space. There are no good reasons for excluding these things from definitions of curriculum as they are, in many cases, the calibrating anchors for curricular trajectories, and key avenues by means of which ethnicity, gender, and class contextualize and are contextualized in curriculum processes.

CONCLUSION

The view of curriculum I have outlined here can be summed up as a number of claims. First, the meanings of instructional events cannot be inferred from course artifacts or teachers' stated intentions. The spatial networks that contextualize such artifacts, actors, and events constitute them, and they emerge over variable but sometimes extended periods of time. Instructional events do not have meaning in isolation; instructional events must be fitted together, accumulated, calibrated, and anchored in relatively extensive and durable systems of activity. Second, contexts are not static encircling frames, but are partly presupposed and partly created through the actions of students, professors, and distant participants like textbook writers, admissions officers, and employers. It is better to talk of "contextualization," a process, than context. It follows, third, that activities like schooling are in large part constituted by struggles over which contextualizations or what kinds of contexts are to hold for participants in particular events, and what kinds are to count in institutional records of those events. Fourth, in this way of thinking, curricula are networks of mutually constraining contextualizations extensive in space and durable over time. They are not reducible to stable bodies of knowledge or skill; nor are they isomorphic with university programs, though in some cases they overlap such programs very closely. Fifth, text artifacts or instructional events like courses, modules, or workshops can be recontextualized to fit into a range of curricula, such that students reading the same book and sitting in the same classroom can be experiencing them in radically different ways as components of different curricular trajectories. Sixth, and finally, contextualizing frames extend beyond the course, even beyond the school, and are anchored, with variable degrees of tightness, to non-school-based activity systems As a result, undergraduate education is regionalized, with some regions being tightly linked to post-graduate labor markets, others loosely connected, and some playing a more or less credentializing role (Collins, 1979).

Things like the nutrition slide (or a whole course composed of such slides) can thus be treated as "content," in the sense Levine (2001) invokes in the quote at the outset of this essay, only by ignoring their variable contextualizations and meanings. This can be easy to do: The wide range of student recontextualizations in the nutrition course, for example, are masked in official accounts by the use of grades which categorize as having performed at different levels of proficiency on the same tasks. However, the meaning of a course like Nutrition, whose ultimate function for most students lies in its contribution to a grade point average, and which is not tightly linked to specific institutional calibrations beyond the university, may not be very important to anyone. This will not be true for other courses. Taylor's (2000) idea of a floating curricular node-scape, or Wallhaus's (2000) suggestion that "conventional" instruction involves "learning objectives ... specified by the institution" while in "e-learning" "learners shop for opportunities that fit their specific needs" are untenable. "Lines of investigation" and "specific needs" are defined in terms of particular systems of activity—jobs, careers, and the like—which serve as calibrating anchors for students' as well as professors' practice. What is changing is not the nature of undergraduate curricular trajectories, but their geographies, the scale at which curricular decisions are being made. There are still objectives, hierarchies, and sequences, but instead of being defined by university-based disciplinary practitioners, they are defined by professional and corporate practices, and instead of being stabilized in particular locations where groups of students meet, they are dispersed across a range of settings.

Is this a bad thing? From a student standpoint, nostalgia for the traditional university may seem largely unwarranted. Disciplinary dominance of curricula is not necessarily any more benign than corporate control (in some areas the two have long overlapped). Reading Powerpoint slides over the Internet for a course you don't care about may not be much worse than sitting in a cavernous auditorium with hundreds of other students listening to someone in the distance read them. The danger is that the equivalence of curriculum with "content" and the focus on idealized curriculum practices rather than careful analysis of what professors and students actually do will waste the opportunity, generated inadvertently by the introduction of instructional technologies, to rethink curriculum and learning in higher education by rethinking the spatial and temporal organization of undergraduate studies and their articulations with worlds outside the university.

ACKNOWLEDGMENT

Thanks to Richard Edwards and Robin Usher for comments that helped me refocus the direction of this paper, and to Liz Barber for critical readings of several late drafts that forced me to give some cohesion to the piece.

NOTES

1. The study also included audio-recordings of the professor working on web materials with instructional designers, audio and video recordings of class sessions, interviews with the instructor and instructional designers over a two-year period, interviews with university administrators and faculty involved in computer-mediated instruction, and analysis of memos, reports, and policy statements produced at the university and the state during the 1990's. None of this material figures in the present chapter.
2. Introductory physics courses can be very large as well. They are large because they are required for students trying to work themselves into the trajectories of science and engineering fields (few students outside the sciences take such courses as electives). Hence professors teaching such courses are under no obligation to make them palatable to students, and may even design the courses to intentionally fail a high percentage of students (Nespor, 1994).

REFERENCES

Bauman, R., & Briggs, C. (1990). Poetics and performance as critical perspectives on language and social life. *Annual Review of Anthropology*, 19, 59-88.

Becker, H., Geer, B., & Hughes, E. (1968). *Making the grade: The academic side of college life.* New York: John Wiley.

Bowker, G., & Star, L. (1999). *Sorting things out.* Cambridge, MA: MIT Press.

Callon, M. (1986). Some elements of a sociology of translation: Domestication of the scallops and the fishermen. In J. Law (Ed.), *Power, action and belief: A new sociology of knowledge?* London: Routledge and Kegan Paul.

Callon, M. (1991). Techno-economic networks and irreversibility. In J. Law (Ed.), *A sociology of monsters: Essays on power, technology and domination.* London & New York: Routledge.

Callon, M., & Law, J. (1997). After the individual in society: Lessons on collectivity from science, technology and society. *Canadian Journal of Sociology, 22*(2), 165–182.

Collins, J. (1996). Socialization to text: Structure and contradiction in schooled literacy. In M. Silverstein & G. Urban (Eds.), *Natural histories of discourse.* Chicago: University of Chicago Press.

Collins, R. (1979). *The credential society.* New York: Academic Press.

Cooper, R. (1998). Assemblage notes. In R Chia (Ed.), *Organized worlds: Explorations in technology and organization.* London: Routledge.

Hanks, W. (1996). *Language and communicative practices.* Boulder, CO: Westview.

Hutchins, E.(1995). *Cognition in the wild.* Cambridge, MA: MIT Press.

Latour, B. (1987). *Science in action.* Cambridge, MA: Harvard University Press.

Latour, B. (1988). *The pasteurization of france.* Cambridge, MA: Harvard University Press.

Law, J. (1994). *Organizing modernity.* Oxford UK: Blackwell.

Law, J., & Hassard, J. (Eds.) (1999). Actor network theory and after. Oxford: Blackwell.

Lee, B. (1997). *Talking heads.* Durham, NC: Duke University Press.

Lemke, J. (1995). *Textual politics.* Washington, DC: Taylor & Francis.

Levine, A. (2001). Privatizing higher education. In H. Levin (Ed.), *Privatizing education.* Boulder, CO: Westview.

Lupton, D. (1996). *Food, the body and the self.* London: Sage.

Massey, D. (1993). Power-geometry and a progressive sense of place. In J. Bird, B. Curtis, T. Putnam, G. Rovertson, & L. Tucker (Eds.), *Mapping the futures: Local cultures, global change.* London and New York: Routledge.

Massey, D. (1996). Masculinity, dualisms, and high technology. In N. Duncan (Ed.), *Bodyspace.* New York: Routledge.

Munn, N. (1992). The cultural anthropology of time: A critical essay." *Annual Review of Anthropology, 21,* 93–123.

Nespor, J. (1994). *Knowledge in motion.* Philadelphia, PA: Falmer.

Olson, D. (1977). From utterance to text: The bias of language in speech and writing. *Harvard Educational Review, 43,* 257–281.

Rosaldo, R. (1989). *Culture and truth.* Boston, MA: Beacon Press.

Sanday, P. R. (1990). *Fraternity gang rape: Sex, brotherhood, and privilege on campus.* New York: New York University Press.

Silverstein, M. (1976). Shifters, linguistic categories and cultural description. In K. Basso & H. Selby (Eds.), *Meaning in anthropology.* Albuquerque: University of New Mexico Press.

Silverstein, M. (1993). Metapragmatic discourse and metapragmatic function. In J. Lucy (Ed.), *Reflexive language: Reported speech and metapragmatics.* Cambridge: Cambridge University Press.

Silverstein, M. (2001). The limits of awareness. In A. Duranti (Ed.), *Linguistic anthropology: A reader.* Malden, MA: Blackwell.

Silverstein, M., & Urban, G. (Eds.) (1996). *Natural histories of discourse.* Chicago: University of Chicago Press.

Smith, D. (1987). *The everyday world as problematic.* Boston, MA: Northeastern University Press.

Smith, M. (2000). *Transnational urbanism: Locating globalization.* Oxford: Blackwell.

Smith, N. (1993). Homeless/global: Scaling places. In J. Bird, B. Curtis, T. Putnam, G. Rovertson, & L. Tucker (Eds.), *Mapping the futures: Local cultures, global change.* London and New York: Routledge.

Star, S.L., & Ruhleder, K. (1996). Steps toward an ecology of infrastructure: Design and access for large information systems. *Information Systems Research, 7*(1), 111–134.

Taylor, M. (2001). Unplanned obsolescence and the new network culture. *Chronicle of Higher Education, 48*(16), Dec. 14. http://chronicle.com/weekly/v48/i16/16b01401.htm.

Twigg, C. (1996). *Academic productivity: The case for instructional software.* A Report from the Broadmoor Roundtable, Colorado Springs, Colorado, July 24–25.

Wallhaus, R. (2000). E-learning: From institutions to providers, from students to learners. In R. Katz & D. Oblinger (Eds.), *The "E" is for everything: E-commerce, e-business, and e-learning in the future of higher education.* San Francisco, CA: Jossey-Bass.

Zack, I. (1998). Masses in classes perplex professors. *Roanoke Times,* 19 August, B1, B3, B3.

CHAPTER 8

JOURNEYS OUT OF PLACE

Nursing Students and Spatiality

Peter Gray

In this chapter, I discuss some of the ways in which nursing students in Scotland experience spatiality in the context of the placement, an aspect of the curriculum with special significance to nursing students. By spatiality, I mean simply the human relationship to spaces, such as lecture theatres, hospitals, or a "room of one's own." I will argue that, as embodied beings, we are inherently spatial and must experience space(s) in some way or another. This experience of space(s) can be enriching, frustrating, useful, or useless for any particular individual. Because spatiality is a global concept that encompasses a number of different topics, spatiality can be a useful way into the study of student experience. In order to analyse the complexity of spatial experience however, I will propose a structural model of that experience which points to it having three "dimensions"—proximity, mobility and possession. These are necessary features of being-in-the-world. Although these dimensions are common to all spatial experience, the nursing student experience of space is particularly rich. It is also of interest in relation to current concerns about nurse recruitment and retention. I will have more to say about the background of nurse education below.

Space, Curriculum, and Learning, pages 109–122

In section I, I outline the theoretical background to the study, and explain what I mean by "experiencing spatiality." Section II presents some examples of what this might mean in relation to empirical data drawn from nursing students. Finally, I conclude by examining whether this way of exploring nursing students' experience of spatiality adds anything to existing knowledge about the student experience generally.

THE EXPERIENCE OF SPATIALITY

Urry (2002) claims that "there is little empirical material" in the study of spatiality, and my study aimed to find out how people (in this case, nursing students) experience spaces in the context of their everyday lives. Although a certain amount of research had been done into the quantitative ways in which people use spaces (Horne, 1999; Hillier, 1996), few previous studies had explored the experience *of* space(s), as distinct from experiences *in* space(s). "Experience" is itself a contested term, but a useful working definition is given by Negt (quoted in Boreham *et al.*, 2002): experiences "are in a certain sense collective moments of dealing with reality which society produces...This is why the term experience always contains a general element which goes beyond the totally individualized, coincidental and purely subjective perception."

Thus, my experience of walking down a street is one of engaging with a socially interpreted but physically substantial reality. In doing so, I conform to assumptions and rules of behaviour (not walking out in front of cars or sleeping on the pavement), mediate my perceptions through cognitive and affective frames ("If I go down here I get to the supermarket"; "I'm fed up with this journey"), and find myself in broad agreement with the actions of others who are also experiencing the space. Exceptional events (someone decides to build a barricade across the street as a protest against globalization) may bring the collective interpretation into question, and perhaps even produce a new interpretation, but the essence of the experience lays in the interplay between its personal and collective aspects. Here, I am interested in the interplay between personal experiences of space and collective understandings of it. In order to examine this interplay in the context of empirical material, a structure is required, and I will now outline the theoretical framework, which was used in the analysis of the data.

Whilst the use of prior theoretical frameworks or assumptions might be frowned upon by "pure" phenomenologists, I follow Schmuttermaier and Schmitt (2001) in arguing that the posing of research questions is itself an imposition upon the data, and that methodological purity is thus illusory. The framework which I propose here is not the only one which might have been used to approach the question of "how nursing students (or any other group) experience space(s)." It would, for example, have been possible to use the affective relationships with spaces that emerged from the data, such as comfort, frustration or anger as a means of structuring participants' responses. As Lakoff (1987) argues, however, most of the ways in which emotions are expressed in language rely on metaphors drawn from physical domains. It is, therefore, worth exploring the experiential basis of embodied spatiality as a prequel to other, more targeted studies.

Heidegger (1962) provides a useful account of how spatiality manifests itself for human beings. It is crucial to note that Heidegger does not see space as an abstract concept (as it is for Newton or Einstein) but as a derivative of being-in-a world and of using equipment (in a wide sense) in a world. Following from this, there are two core concepts that enable us to have any experience of space at all. These are the complementary concepts of "distance" and "dis-stancing" or pre-reflective concern. The two concepts are not separate but complementary. Distance, in its most basic sense, can only be experienced because things, and especially "equipment" and "others" are within our range of concerns—concern tells us what or whom we are dis-stanced from. To be concerned with something is thus to have a relationship to it, a relationship which I refer to as proximity. Proximity here does not mean co-presence but covers any relationship of dis-stancing, whether the thing in question is physically near or far. We are all in proximity to some thing(s), in varying degrees, and whilst this dimension is the most basic feature of our experience of spatiality, proximity does not encompass the possibility of action over distance, which I therefore describe as mobility (Albertsen & Diken, 2001). As with proximity, "mobility" implies a range of mobilities, from immobility to complete mobility across a spectrum. As Urry (2002) points out, mobility is highly differentiated and strongly related to social capital. It also underlies many areas of private and public life. Constraints upon mobility (asylum, prisons, disability) and enhancements of mobility (transport infrastructure, trade agreements, transferable skills) form major elements of policy.

The two concepts of proximity and mobility thus embrace the possibility of distance, differentiation and action over a distance, but the interaction of time and space is not fully captured either in proximity or mobility, other than in the concept of speed (Plumb, 1999). As Massey (1992) points out, time and space cannot be considered in isolation from each other, and at an everyday level it seems obvious that spending time anywhere increases familiarity (or boredom) with that particular space. Heidegger (1962) suggests that there cannot be *any* kind of meaningful spatiality without a temporal dimension to "stretch" it out in every direction. By "haunting" (a) space, we become familiar with it, and our emotional relationship with that space develops (De Certeau, 1984). That relationship may, of course, develop into anger, frustration, delight or indifference, but even indifference to a particular space constitutes an affective relationship, which we cannot avoid having (Malpas, 1999). Furthermore, our temporal relationship to a space always involves others, even if it involves their exclusion, or their absence as memories or "ghosts." As an analytic category to gather these relationships together, I use the term "possession." Possession does not imply a static relationship but is a variable sense that takes account of the power-relations, which permeate any space Foucault (1984). In effect, possession is the bridge between personal and collective experiences of spatiality, as in my discussion of the experience of the street above. My relationship with the space of the street is conditioned by the aggregate of power relations, which permeate it, whether through traffic regulations, the presence of menacing gangs or my (hypothetical) role as a local councillor.

The hypothesis here is therefore that spatial experience might be usefully considered as having three dimensions—proximity, mobility and possession. By using these analytically, the experience of spatiality can be explored in greater depth. This study draws upon a particular set of spatial experiences that of nursing students, for reasons which I will now explain.

Although it is concerned with the education and training of the largest single workforce in the UK, nurse education has never had "a place of its own." It has, by various accounts, a history of subordination to the medical profession (Davies, 1995), a fundamental uncertainty about its own image (Chiarella, 2002) and a turbulent relationship to health policy (Rafferty, 1996). The Project 2000 initiative (UKCC, 1986), however, had the effect of placing nurse education firmly within the remit of higher education. One of the main features of Project 2000 was that "student nurses" became "nursing students." They gained supernumerary status, and were, therefore, no longer rostered members of staff on wards or elsewhere. However, this did not mean that their education became purely theoretical. It was, and still is, considered necessary for nurses to engage in practice placements in a variety of settings, in order to apply their academic learning to practice and to have that practice assessed. Recently there have been moves to reinforce the curricular importance of practice in the placement situation, albeit supported by evidence drawn from the expanding research base of nursing (UKCC, 1999). There has been some discussion of the placement experience in the literature (e.g., Cope *et al.*, 2000; May *et al.*, 1997), but this has largely ignored the spatial aspects of that experience. The study on which this chapter is based attempts to address the lack of research in this area, using the three-dimensional matrix for the analysis of spatial experience, which I described above. In the next section I will discuss the empirical data and show how this matrix might be used to analyse some participant statements.

NURSE EDUCATION AND THE SPACE OF PLACEMENTS

As nurse education is experiential as well as academic, nurse students are required to make frequent transitions from one kind of space to another. Some of their learning takes place within academic institutions, some of it in their own homes, and some of it in placement settings. It is the students' experience of the spatiality of placements that is the focus of this section. The study did not attempt to relate this experience directly to learning outcomes for several reasons. Firstly, the experience of spaces is highly variable and creates difficulties in sampling and control group selection. There was, however, an element of comparison in that two rather different academic-institutional settings were involved. One was a medium sized university campus in a well-populated lowland area of Scotland, whilst the other was a former nursing college (now attached to the same university) in a rural area of low population density and large geographical area. Secondly, the question of learning outcomes is itself contested, with advocates of competence-based approaches competing with those who advocate a more holistic approach (Benner, 2000). Given its limited time and resources, the study was concerned to establish whether the spatial analysis of experience along the dimensions suggested above would be useful in illuminating problematic areas of the nursing student experience as a whole. All the students in the study were mature students, many of who had already been nurses for some time. What I wished to explore were the ways in which the participants in the course of the educational process experienced these physical spaces. Rather than influencing learning outcomes, these experiences may influence the morale and therefore the retention rates of nursing students, both of which are crucial to fulfilling the current demands of health care policy.

May *et al.* (1997) define the placement as "community-based or clinical settings in which the student is placed for experiential learning purposes." The intention of the placement is to integrate theory and practice and to begin the important process of professional socialization (Gray & Smith, 1999). As I will show, it is perceived by the students to fulfil these functions, but it does so at the cost of (in some cases) great personal effort. The placement is a pre-existing world into which the student is "thrown" (Heidegger, 1962), meaning that it is always already "up and running" when the student arrives (and, of course, it continues after her departure). In the placement setting, the student is widening her "range of concerns." She is brought into a proximity relationship with previously unfamiliar objects and others. May *et al* (1997) draw attention to the benefits of continuity where students have been able to return to a previous placement, which suggests that the process of gaining possession of the space of the placement is part of the overall process of professional socialization. In terms of mobility, the student, although thrown into a pre-existing situation, has exercised mobility simply by getting to work. In the institutional setting under discussion, participation in placements called for an especially high level of mobility, due to the geography of the largely rural area across which the students were dispersed. I will return to this topic below.

In order to explore the spatial experience of placements, it seemed probable that the aims of the study would be best served by using interview methods, following Hollway and Jefferson's (2000) critique of the limitations of questionnaires and survey techniques in investigating the complexities of everyday life. I explained to the participants in my preamble that I was interested simply in the "where" of their study habits and learning experiences, but once the interviews had commenced, students moved easily into talking about the "how" and "why" of particular spaces.

The interviews allowed the students to talk about any relationship with space in the context of the nurse education process as they saw it. Of course, their identities within this frame were not unproblematic. Many of the students had been professionally involved in nursing prior to re-entering full-time education, and brought with them a strong sense of their working identity. They were nurses first and students second. Their nursing identities became problematic for them when they encountered resistance to their performance of those identities. Chiarella (2002) suggests that there are at least five "stories" which are told about nurses, or which nurses tell about themselves: the domestic worker; the doctor's handmaiden; the subordinate professional; the ministering angel; the autonomous professional. These stories are told for all sorts of reasons and nurses frequently find themselves in roles, which are framed by more than one story at the same time. Thus, nursing students are socialized into roles whose definition, or discursive location, is uncertain. One way of examining this uncertainty is to see it as a question of proximity and thus to ask what participants have within their range of concerns in a given situation. These proximities act to locate the participants both within and across the boundaries of the specific situation. One of the important proximities with which the participants were concerned was their proximity to patients. Because the placement was an opportunity to exercise their nursing identities in a specific spatial situation, anything that interfered with this proximity to patients affected their sense of possession of that space. One student expressed this sense of (dis)possession: "the common foundation course, it was appalling, just ridiculous, I spent six weeks on a health promotion placement... As far as I'm concerned, it had nothing to do with nursing... it was a question of not getting enough appropriate experience when we're on placements... it would be good if there were beds (to make) because that would mean there might be a chance of a patient being there, on the wards" (Student G).

For the student in the next extract, the lack of regular proximity to other students or tutors during her placement (caused partly by the unplanned absence of a fellow student who was due to share accommodation) was highly problematic:

> "I had to have a placement that was quite far away, and I was by myself, and it was very lonely, horrible, there was no contact, although some of the staff you could speak to, but you couldn't speak to (anyone else), so you had like contact deprivation almost, you had four walls and you're bouncing off them, all you've got to do is your essay to concentrate on, you can only do that for so many hours without becoming nutty (laughs)." (Student S)

The student to be at the furthest extreme of proximity to the university, being only just within its range of concerns, senses the placement. Only the essay acts as a link to the institution, whilst the process of bringing the placement setting fully into one's range of concerns cannot always be managed in the time available. The metaphorical "bouncing off (walls)" mediates the link between the spatial, the cognitive and the social aspects of the situation (Lakoff, 1987). The space of the placement is experienced neither as one's workplace, nor as one's place of learning, but as an "in-between" space in which one's identity cannot be fully asserted. In terms of mobility, it is the student who in this case is highly mobile, whilst the institution is not. The latter's perceived lack of proximity is not countered by institutional mobility. The university does not reach out to the student over distance. In addition to the proximity and possession issues around placements, this extract also hints at the mobility issues faced by nurse students. The assumption here is that physical and professional mobility are an increasing attribute of the workforce in general and nurses in particular. This assumption can be traced to its origins in embodied physical mobility and the relationship with cognitive processes.

Lakoff and Johnson (1999) explain that many of the conceptual-metaphorical structures that we use to deal with ideas (e.g., to "bounce them around") are based on our embodied experience of spatial action in the physical world. Mobility in its basic manifestations can be experienced as pleasurable or unpleasant depending on the power relations that apply in the situation. The same terrain can be covered as a country walk or as a forced march, providing a very different experience in each case. One of the questions which originally inspired the study was "why does walking help thinking?," and this extract begins to capture the relationship: "I really like my time walking, I'd just come out of the (b)uilding and walk, I enjoy the peace it gives me, the time to unwind, and just reflect on things, or if you've just come out of something and your head is buzzing... so I enjoy my walk and the benefits of physical activity as well..." (Student A). Walking, in this case, is a good example of the intertwining of the spatial, the metaphorical and the temporal—moving through physical space in a way, which enables concentration on things other than the activity itself. The above extract continues:

> "It started off, I felt very apprehensive because *it was such a big step* and it was new and you felt like this new kid on the block, even though you knew there were other people, there were feelings of apprehension, now you just concentrate, I really enjoy the walk, all that wildlife, there's a pleasurable sense as you walk along, of the smells and the flowers and seeing it turn all green and lovely which is nice, and then you walk across the bridge and that's you *in* and I'm concentrating on *what's ahead of me*." (Student A, emphasis added)

The interesting points here are the shift from the physical spatiality of walking to the metaphorical spatiality of the "big step," the shift from a weak to a strong sense of proximity (from "outside" to being "in") and a concurrent increase in the sense of possession ("now you just concentrate...") as the "feelings of apprehension," which accompany the move into a new space, are overcome. These are, of course, social processes and could be studied as such, but the point is that they are mediated by the experience of embodied spatial activity. I am not arguing here for a separation of social and spatial activity, but rather for the opposite, an acknowledgement that social activity cannot be divorced from its spatial locations (Baldry, 1999).

A rather different example of the connection between activity and location is given in the next extract. In the geographical area, which was served by one of the institutions in the study, the wide spatial dispersion of students on placements had significant effects, which were expressed in the following extract:

> "(The journey") was ghastly, it would have been fine if you could see something but driving up the loch side in the dark in the middle of winter is like, 'what am I doing? And why am I putting myself at this risk?' And my kids of course, there were lots of doubts about whether this is remotely sane, and I do... one of my big things is, single parent families in particular, should somehow be excluded from this nonsense, because if I die the kids have had it and I think that is a huge risk to take." (Student C)

The point here is that the placement experience is not isolated in time and space, but is tied into multiple networks involving, in this case, parental help with childcare and a related bundle of organizational arrangements, which themselves contributed to the learning experience of the placement, as the second part of the above quotation suggests:

> "...but having done it, there were lots of positive things came out of it, and it was a huge learning curve, and staying in bed and breakfast was brilliant, but

> masses of organization and planning, getting your children looked after, pack them off, pack their week's things, 'when have they got piano? When have they got ballet?' Lots of planning, you needed amazing organizational skills, but once you had done it, what a sense of achievement!" (Student C)

The institution, in this case, was simply "not there" for, or not in sufficiently close proximity to, the student, in its failure to take the domestic responsibilities of the student into consideration. The fact that the student regarded the overall learning outcome as positive reflects a pragmatic and self-sufficient approach to the nurse education process, one, which is conditioned by the identity "stories", referred to by Chiarella (2002).

Similarly, the issue of temporary accommodation, which students used whilst on placement, formed part of this experience of low institutional proximity. Most of this accommodation was not attached to placement settings, but consisted of bed-and-breakfast rooms intended for tourists. The process of obtaining accommodation was thus haphazard and dependent on local demand, and students received little institutional support for this aspect of the placement process, another of the ways in which the university's marginal proximity manifested itself. Their mobility, their action over distance, was thus dependent on a certain level of autonomy and initiative: "...really it is finding accommodation that's the big problem, because in the winter a lot of the places are closed down and in the summer all the prices are horrendous and you're working early shifts and things and a lot of the B and Bs (bed-and-breakfasts) you don't get any breakfast until 8 o'clock or whatever, so yes you can [find that a problem]" (Student D).

These problems might be expected, given the low population density of the area, but there were other issues around the nature of the accommodation and its spatial arrangements. Studying as an activity was required to be "portable," but is also conditional on having a certain relationship to the available space(s), as the following extract explains:

> "When I've been away, it's always been B and B accommodation, there is just this feeling that it's not entirely your own space, because you're living in somebody else's house, and it's not your chosen surroundings... so far I haven't had television in my room, but there has been television in my B and B, and it's been a distraction. I am actually staying in halls (of residence)... this time instead of a B and B and I'm finding that better, because they've got the set-up, when you're in somebody else's house... you don't often have a desk, whereas this one has and it's a bit more equipped for students, so it's a bit better this time round... whereas in halls everybody shares, so it's a bit more like it's your territory." (Student L)

Again, the extract highlights the need for student accommodation to be fit for its purpose, but it also reveals the importance of possession as a power relationship, the sense of being in control over a space, or of having one's own "territory," which conforms to the generic standard appropriate to its purpose. Student rooms in halls of residence form one such standard, and offices generally form another. In the extract above, there is a sense of reduced possession because the participant has to study in a space, which is geared towards relaxation and transient occupation. Thus "studying" is constructed as a portable activity, which can be carried on in a variety of locations. As we have seen, the students' perceptions of the effectiveness of study in unsuitable locations show that the portability of "studying" cannot be taken for granted.

The next extract also reveals that 'fitness for purpose' is linked to gaining a sense of possession:

> "It depends on the placement. When I was up in Wick, it was a health promotion placement, and I took all my books up with me, I even took my word processor, and I did a lot of work, I didn't want to communicate with anybody, I just shut the door, the accommodation at Wick was very good, it was purpose-built for the nursing staff, we had our own space and shared showers and so on, but at this place I've just been so tired at the end of the day that after the first week I thought, there's no way I'm going to study, I'm not finished till half-past eight at night, I don't get back till quarter to nine and by the time I've had something to eat, you just don't feel like it, there's no point, you're just not taking anything in..." (Student E)

On the one hand, the workload of the placement sometimes meant that students were unable to focus on study in general and on essays in particular during their off-duty time. On the other hand, the placement provided opportunities for some students, usually those with children, to spend extra time studying and writing, since they were relieved of their domestic workload. This is the converse of the problematization of "home" as a study space, and one of the unexpected benefits of the placement here is the relative luxury (for students with families) of being looked after in bed-and-breakfast accommodation—no bed-making, cooking or washing up. This was not a unique experience, as one of their tutors pointed out: "yes, they do sometimes (enjoy it), to get some quiet evenings and time to be entirely a student, for a few weeks and not to have to worry about shopping and picking up the kids..." (Tutor J). This highlights the constraints which often render "home" a contested space for students, and which require a high level of informal support, in cases where the institution fails to provide suitable arrangements:

> "It doesn't create a problem (with domestic arrangements), because I'm pretty pragmatic about these sort of things, but it does annoy me that I've had to go away, because I should have had a placement at home, but there

> weren't enough placements so five of us got moved to the outlying districts, so there could be an element of annoyance there, but it doesn't create particular difficulties for me with my home life, because my family are behind me... everybody has difficulties, but it's not particularly because I'm away." (Student E)

This "pragmatic" approach can again be seen as expressive of the "nurse as domestic worker" story which, as Chiarella (2002) shows, resulted in poor living conditions and a harsh disciplinary regime for many nurses until relatively recently. As part of the experience of spatiality, accommodation problems can generate a diminished sense of possession. Equally, as in one of the extracts above, the experience of tourist accommodation can generate a heightened sense of possession through the "luxury" of relief from domestic chores and responsibilities. This, however, is not a planned benefit, although it illustrates the way in which possession is a contingent interaction between individuals and spaces.

The marginal proximity of the university while on placements was thus a factor in students' perceptions of autonomy and therefore their sense of possession in the placement itself. The high level of mobility which the students generally delivered to the institution reflects the "autonomous professional" self-image of nursing discussed by Chiarella (2002), an image which is reflected in the following extract. "For me personally, the overwhelming impression I get is that they're (tutors) far too busy, and it's just the usual thing that you don't want to appear incompetent or a nuisance, or particularly annoying, or *rocking the boat*, or [to be] pestering them, so I just feel I've got to be confident enough to *ride this out* for myself" (Student E, emphasis added). The nautical metaphors here ("rocking the boat... ride this out") convey a sense of individual resilience within the context of a close-knit group. As the participant makes clear, this desire not to appear incompetent is "the usual thing", as indeed it was for all the participants who expressed a view on the topic. There is also an element here of the "ministering angel" image which in Chiarella's view has frequently resulted in poor pay and conditions for nurses.

The experience of placements, then, contained a number of contradictions, particularly for those students with young families. On the one hand, childcare arrangements had to be made and various forms of support organized, but on the other hand, the placement provided opportunities for uninterrupted study time, through freedom from domestic chores. Despite the nominal supernumerary status of the students, placements sometimes involved exhausting shift work, which left little energy for study at the end of a 12-hour working day. The advantages of studying away from home were therefore sometimes outweighed by the disadvantage of having to be part of a working team. This in itself was a reflection of the students' own ambivalence about their supernumerary status. On one hand, this was intended to prevent them being used as "domestic labour" (Chiarella, 2002) on the wards, but on the other, it prevented them "getting stuck in" as one of them put it. This applied particularly to those with previous experience.

The question of possession in a placement situation was thus an interesting one. Students were keen to fit themselves into, or to come into proximity with, pre-existing teams and thus to gain a sense of possession of the particular setting involved. Thus they negotiated and gained possession, for example, by making sure that they learned how to find essential items, itself a form of mobility, or by participating in tedious routine activities in order to demonstrate a willingness to share in the work. Possession is more than "integration into a team," a "community of practice" (Wenger, 1998) or an institutional culture. Possession, as I conceptualize it here, also takes account of the spatial experience that brings the student to her location within the team, community of practice or culture.

CONCLUSION

What I have tried to show in this chapter is that spatiality is a three-dimensional experience which integrates the three components of proximity, mobility and possession, and that an enhanced understanding of this spatial experience contributes to the knowledge base of nurse education. The proximity dimension is relevant because it helps to clarify the ways in which different spaces focus or disperse the concerns of the students, both professional and personal. Spaces are experienced both collectively and individually, and in this case the curriculum of nurse education is realized within the boundaries of specific spatial experiences.

The mobility that was required of the students in fulfilling the curricular demands of the placement is founded on a wider social sense of mobility, which has been problematized in the literature of social geography (Urry, 2002; Albertsen & Diken, 2001; Adams, 1999). Its use here as an analytical tool has challenged the assumption that nurses, irrespective of their differentiation along lines of gender, age or domestic situation, were willing to engage in unlimited travel, an assumption built into current policy discourses of "flexibility" and "occupational mobility" (HM Treasury, 2002; UKCC, 1999).

In general, students felt proximity both to the educational institution and to the various work settings through which they had passed on the way to their current state. It is important to analyse this multiple sense of proximity because one of the long-term problems of nurse education has been its uneasy relationship with the various agencies involved in the provision of education and training (Rafferty, 1996; Davies, 1995). Nursing has historically been a contested activity, as practice or academic discipline, art or science, care-work or the application of scientific knowledge (Benner, 2000). These perceptions of nursing are actively contested in current discourses, but what is rarely considered is the spatial element of this contestation. A simplistic division between theory and practice is unhelpful here. Nursing is currently positioned not only between the *discourses* but also between the *spaces* of education and of health care. This is not to argue for a return to the stereotypical nurse-training model of a ward-based practical apprenticeship. It requires a more nuanced consideration of the spatial issues around nurse education. Writers in the field of spatiality generally draw attention to the importance of spatiality as an issue, but what is required is further analysis of the global concept of spatiality into more manageable components. This chapter has attempted to provide such an analysis.

ACKNOWLEDGEMENTS

The work on which this chapter is based was supported by an ESRC postgraduate studentship no.R00429934171. I would like to thank Richard Edwards for helpful comments on earlier versions.

REFERENCES

Adams, J. (1999). The social implications of hypermobility. In *Project on environmentally sustainable transport (EST), the economic and social implications of sustainable transportation—Proceedings of the Ottawa workshop.* Paris: OECD.

Albertsen, N., & Diken, B. (2001). *Mobility, justification & the city.* Lancaster: Dept. of Sociology, University of Lancaster on-line paper accessed 24/1/02 at http://www.comp.lancs.ac.uk/sociology/soc082bd.html.

Bachelard, G. (1964). *The poetics of space.* Boston: Beacon.

Baldry, C. (1999). Space—the final frontier. *Sociology, 33*(3), 535–554.

Benner, P. (2000). The quest for control and the possibilities of care. In M. Wrathall & J. Malpas (Eds.), *Heidegger, coping and cognitive science: Essays in honor of Hubert L.Dreyfus: Volume 2.* Cambridge, MA: MIT Press.

Boreham, N., Samurçay, R., & Fischer, M. (Eds.) (2002). *Work process knowledge.* London: Routledge.

Chiarella, M. (2002). *The legal and professional status of nurses.* Edinburgh: Churchill Livingstone.

Cope, P., Cuthbertson, P., & Stoddart, B. (2000). Situated learning in the practice placement. *Journal of Advanced Nursing, 31*(4), 850–856.

Davies, C. (1995). *Gender and the professional predicament in nursing.* Buckingham: Open University Press.

De Certeau, M. (1984). *The practice of everyday life.* Berkeley: UCP.

Foucault, M.(1984). *The foucault reader.* New York: Pantheon Books.

Gray, M., & Smith, L. (1999). The professional socialisation of diploma of higher education in nursing students (Project 2000): A longitudinal qualitative study. *Journal of Advanced Nursing, 29*(3), 639–647.

HM Treasury. (2002). *Securing our future health: Taking a long-term view—an independent review by Derek Wanless.* London: HM Treasury.

Heidegger, M. (1962). *Being and time.* New York: Harper Row.

Hillier, B. (1996). *Space is the machine.* Cambridge: Cambridge University Press.

Hollway, W., & Jefferson, T. (2000). *Doing qualitative research differently.* London: Sage.

Horne, S. (1999). *The classroom environment and its effects on the practice of teachers.* PhD Thesis, Dept. of Design Studies, Goldsmiths College, University of London, unpublished.

Lakoff, G. (1987). *Women, fire and dangerous things.* Chicago: University of Chicago Press.

Lakoff, G., & Johnson, M. (1999). *Philosophy in the flesh: The embodied mind and its challenge to western thought.* New York: Basic Books.

Malpas, J. (1999). *Place and experience: A philosophical topography.* Cambridge: Cambridge University Press.

Massey, D. (1992). Politics and space/time. *New Left Review,* 196, 65–84.

Massey, D. (1994). *Space, place and gender.* Cambridge: Polity Press.

May, N., Veitch, L., McIntosh, J., & Alexander, M. (1997). *Preparation for practice: Evaluation of nurse and midwife education in Scotland: 1992 programmes.* Glasgow: Dept. of Nursing and Community Health, Glasgow Caledonian University.

Plumb, D. (1999). Adult education in a world 'on speed'. *Studies in Continuing Education, 21*(2), 141–161.

Rafferty, A. (1996). *The politics of nursing knowledge.* London: Routledge.

Schmuttermaier, J., & Schmitt, D. (2001). Smoke and mirrors: Modernist illusions in the quantitative versus qualitative research debate. *Sociological Research Online, 6*(2), <http://www.socresonline.org.uk/6/2/schmuttermaier.html> accessed 23/4/02 .

UKCC. (1986). *A new preparation for practice.* London: UKCC.

UKCC. (1999). *Fitness for practice.* London, UKCC.

Urry, J. (2002). Mobility and proximity. *Sociology, 36*(2), 255–274.

Wenger, E. (1998). *Communities of practice: Learning, meaning, and identity.* Cambridge: Cambridge University Press.

CHAPTER 9

A GEOGRAPHY OF POWER, IDENTITY, AND DIFFERENCE IN ADULT EDUCATION CURRICULUM PRACTICE

Arthur L. Wilson and Ronald M. Cervero

Why would it matter "where" we put our educational programs? After all, isn't "place" just empty space, a proscenium arch framing but not much influencing human action? Why should we care where learning and education occur as long as we get the content right, use appropriate methods, employ good teachers, produce the right results for our stakeholders? For example, why could it possibly matter where we locate a continuing education program for a profession? Does it matter whether such adult education were placed on a college campus or located at a luxury business resort or even transmitted "through" cyberspace via distance technologies? Does where we put our adult education programs have anything to do with what we produce with them? Does place matter?

We use this chapter to ask "why and by what means do social beings invest places ...with social power; and how and for what purposes is that power then deployed and used across a highly differentiated system of interlinked places?" (Harvey, 1992). Our answer is that place matters because where we place our programs contributes to producing difference in two senses: constructing identity and exercising power. If place is

Space, Curriculum, and Learning, pages 123–138

involved in the production of identity and power differences, we then ask what place has to do with the production of justice (Harvey, 1996). We conclude with restating earlier arguments (Cervero & Wilson, 1994, 2001) that we understand adult education as a socio-political practice of producing and reproducing interests and power and understand adult educators as knowledge-power brokers who choose sides in distributing knowledge and power. The chapter is organized in two parts. In the first we present our understanding of how place contributes to producing identity, power, difference and how we could assess such production. In the second we provide an example of how place contributes to such production in a continuing education program.

PRODUCING JUSTICE AND DIFFERENCE

Harvey (1996) argues that the "fluxes and flows" of socially constructed meaning and action are also a function of the "gritty" material realities[1] in which access to and denial of material and symbolic benefits become emplotted and unfold. Understanding how these relatively enduring systems of value, institutional ideologies, and social practices get constructed and changed is the focus of this chapter. "We need not only to understand but also to create permanences—organization, institutions, doctrines, programs, formalized structures, and the like—in order to change anything in any kind of meaningful or directed way" (Harvey, 1996). Understanding the social construction of space is necessary for understanding how difference gets produced in specific places (the political question) and for what purposes (the ethical question). Developing a "cartographic imagination" (Edwards, Cervero, Clarke, Morgan-Klein, Usher, & Wilson, 2002) helps us to understand how geographies of power work.

THE "REPLACING" OF SPACE AND PLACE IN SOCIAL ANALYSIS

To begin to see geographies of power, we must have a different understanding of what space is and how it works. In a stream of analysis focused on "reasserting space" into social analysis (Bird, Curtis, Putnam, Robertson, & Tickner, 1992; Edwards & Usher, 2000; Friedland & Boden, 1994; Lefebrve, 1974; Soja, 1989), it is becoming increasingly clear that not only are knowledge and power interconnected, but that knowledge, power, and space/place closely intertwine to frame our social practices. Foucault (1980) has commented that "a whole history remains to be written of spaces—which would at the same time be the history of powers," noting that space is fundamental in community life and central to the exercise of power. By linking space, knowledge and power, we take Foucault to mean that structures of meaning and social practices which are both derivative and constitutive of the production and use of power are themselves implicated and produced recursively in specific places. While the thing we call "space" may begin as a physical construct, "the *organization* and *meaning* of space is a product of social translation, transformation, and experience" (Soja,1989). Important for understanding the role of place in producing power in adult education is that "space is produced and reproduced and thus represents the site and outcome of social, political, and economic struggle" (Keith & Pile, 1993).

Most directly then, space does more than provide the settings for social interaction (Giddens, 1984). It is not simply some backdrop, stage, or empty container of social interaction; it itself is a fundamental constituent of knowledge/power regimes. Harvey (1992) has noted that "there are real geographies of social action, real as well as metaphorical territories and spaces of power" in order to ask what is the relationship between place and power and to what ends is such power then used. In order to understand adult education and lifelong learning, we have to map geographies of them as sets of social practices, a human geography in which power and difference are created, enacted, altered. Creating such geographies means we have to come to understand how space is represented, how its meanings are produced, how identities are formed in relation to specific places, who produces and benefits from all this production—in brief, how difference gets constructed and produced. Space, then, is not anything but rather a active constituent of social life.[2] Such insight is necessary to understand the relationships between place and power in the production of meaning, identity, and difference in adult education curriculum practice.

In an attempt to embody Harvey's (1996) "struggle to find foundational concepts for the human imaginary to contemplate our embeddedness in space, time, nature, and place," we turn briefly to his notions of examining "permanences" and creating processes of valuation as a way of "seeing" the power of place in the production of difference. Harvey acknowledges but challenges the intellectual vogue of reducing everything to fluxes and flows of socially constructed, situationally variable idiosyncratic meanings and interactions. In challenging socially constructed understandings, he does not, however, revert to espousing simple foundational beliefs, as is common in popular and government discourses. He argues that permanences "surround us ... [and] help solidify and give meaning to our lives ... while its formally true that everything can be reduced to flows ... we are in daily practice surrounded by things, institutions, discourses, and even states of mind of such relative permanence and power that it would be foolish not to acknowledge those evident qualities" (1996)[3]. Harvey seeks to situate us in the actual conditions in which we find ourselves, in order both to understand and alter these relatively enduring structures of meanings, social practices, and institutions. To do this, he (1996) invokes the critical role of values because "meaningful political action . . . cannot proceed without some embedded notion of values." Yet he undermines that move immediately by saying that any appeal to presumed permanent values to inspire our collective action is highly suspect.

Seemingly, then, we're stuck, with "no place" to go. The world is largely the result of fluxes and flows fostering instability, yes; but, no, there are relatively enduring, patterning structures which can be both helpful and dangerous; and yet there can be no appeal to enduring sets of Archimedean-like values to constitute and direct meaningful action. In a move now fairly standard to exhorting critique,[4] "The answer to these difficulties ... is to replace the fixed idea of 'values' with an understanding of 'processes of valuation.' Once we come to appreciate how such processes operate, then we can also better understand how and why certain kinds of 'permanence' get constructed in particular places and times so as to form dominant social values to which most people willingly subscribe" (Harvey, 1996).

Why is such critique necessary? "The task of critical analysis is not, surely, to prove the impossibility of foundational beliefs (or truths), but to find a more plausible and adequate basis for the foundational beliefs that make interpretation and political action meaningful, creative, and possible" (Harvey 1996). Surely, though, we must tire soon of this constant invocation of critique. Perhaps not just yet, however, for the stakes remain high. The deep political dimensions and ethical consequences of our work are too often too simply disguised as discipline techniques. The practice of education cannot be viewed solely in technical or procedural terms, as is the wont of so many educators, theorist and practitioner alike, because there is no neutral or innocent place from which to practice adult education (Cervero & Wilson, 2001). Harvey's call for plausibility is for grounding the politically pitched and ethically attuned action required to address the dilemmas and demands of our work. Thus the politics of practice cannot be divorced from the ethics of its consequence. Hence the necessity of critique as some small hope for challenging the permanences in which our educational practices and we are embedded.

So as adult educators we must choose. But which is the right side? How do we know the right side when we see it? Indeed, is there a right side? Along with many others in the intellectual world of adult education, we too have invoked images of social justice and democratic practices in adult education (Cervero & Wilson, 1994, 2001), which by building on American democratic traditions ranging from Dewey to Apple we have labeled "substantively democratic participation." Such an invocation to care about the practice of adult education in a certain way attempts to align itself with the broader question of the "justness" of educational work. Critics (Sork, 1996) have questioned what we mean, saying this invocation to ground justice in democratic practices is not new in American adult education or asking how one ethical principle could comprehensively represent practice. Among the many possible responses, the relevant one here is to ask just how we really understand such ethical principles.

"Is it possible ever to talk about justice as anything other than a contested effect of power within a particular place at a given time?" (Harvey, 1996). In response he (1996) proposes that, because there is no universally agreed upon conception, we should understand justice as a "'family' of meanings which can be understood only through the way each is embedded in a particular language game." Here he argues for a sort of "standpoint" concept of justice, in a way eliding into postmodern relativism. He finds such relativism less troubling than others (forever caught up in *whose* foundationalism is right) because such "discourses are expressions of social power" (ibid.) in terms of who can construct justice and how it is practiced. This is important because, even though concepts of justice are socially constructed within specific historical, geographical conditions, "once constituted, the trace of a particular discursive conception of justice ... becomes an objective fact that embraces everyone within its compass" (ibid.) In a word, a "permanence" that structures relations. Now we see why Harvey believes "processes of valuation" are crucial to the construction of just difference. He (1996) wants "to make it possible to use the mobilizing powers of the quest for justice not as a mere matter of pragmatism or opportunism, but as a principled way in which to talk about the need to regulate human relationships and our collective endeavors so as to achieve a particular set of goals under a given set of ecological, historical, and geographical conditions." It is not the critique of foundationalism that is important, but the production of meaningful political action based on a critically evolving determination of justice. There are spatial dimensions to such practices.

PLACING POWER AND IDENTITY: LOCATING DIFFERENCE

Processes of valuation are thus crucial to the construction of the political and ethical practice of education. What does that have to do with the question of place and curriculum practice in adult education? As a last little bit of intellectual architecture we argue that place is involved in the production of identity and power and thus difference. The question of identity has, of course, a long history. More recent attempts (Bird et al., 1992; Foucault, 1980; Friedland & Boden ,1994) have moved steadily away from the traditional tenure of developmental (read pyschologized) interpretations to more structural analyses. For example, Keith and Pile (1993) reject synchronic or essentialist approaches to put forth notions of identity as becoming or process. What makes such analyses poststructuralist as well as postmodernist is the premium placed on the creation of subjectivity and difference as definers of identity: "Any articulation of identity ... is only momentarily complete.... In such a fragile world of identity formation and object formation, political subjects are articulated through moments of closure that create subjects as surfaces of inscription ... invariably incomplete" (Keith & Pile, 1993). Thus a significant poststructuralist question is the debate between a sanctified cogito and an inscribed subjectivity.

This poststructuralist concern with the inscription of subjectivities is also profoundly locational: identity and location are inseparable, thus inviting a newer question of "locating a politics of identity" (Keith & Pile, 1993). As Bondi (1993) has put it, the postmodern twist of "who am I," with its persistently pesky political connotations, has increasingly become also a locational question of "where am I." Or, paraphrasing Walter Benjamin, knowing oneself is "an exercise in mapping where one stands" (Keith & Pile, 1993). So we argue that to understand the complex dimensions of human interaction, we have to recognize a "locality" to them, a "placeness" that is both constituted as well as constituting (Harvey, 1993), or as Giddens (1979) would say, both outcome and medium. Difference and location combine to define identity within a relational field: "there is no identity outside of its context" (Keith & Pile, 1993) or as Edwards and Usher (2000) put it, "identity is secured through location and locating practices." The impossibility of identity without context introduces a second way of understanding the construction of difference, for Keith and Pile (1993) argue that the "place of politics" cannot meaningfully be separated from the "politics of place." The importance of place in producing power relations is not a new insight, merely an overlooked one (Soja, 1989). Lefebrve (1974) has demonstrated that place is not simply a material object but rather has always been intricately involved in the production of power. As Foucault (1984) has remarked, "space is fundamental in any form of communal life; space is fundamental in any exercise of power." Following Harvey (1992, 1996), we see the need to map geographies of adult education as sets of social practices and power relations; part of such mapping requires analyzing the relationships between location and identity formation "because location is about the exercise of power" (Edwards & Usher, 2000) and identity is relationally constructed through and in place. So difference is defined both in the poststructuralist sense of the construction of subjectivities and in the critical sense of understanding the production of relations of power. Producing identity and power are of course crucial to the construction of difference.

LOCATING IDENTITY, POWER AND DIFFERENCE IN AND THROUGH ADULT EDUCATION CURRICULUM PRACTICE

Here is where we are. We have to see that place cannot be an ignored aspect of plotting geographies of power, that espousing foundational values will not rescue us from the indecision of relativism, and that the meaningful construction of action requires continuing processes of critical evaluation: "a political movement has to make its choice and not reserve its judgment" (Harvey, 1996). Thus the construction/alteration of permanences are tied to the questions of justice. Differences, defined relationally in terms of identity and power, are socio-politically located in specific historical, geographical, material places. We must question their construction (the political question) and who is served and in what ways by that construction (the ethical question).

We turn now to an interrogation illustrating what Harvey means by processes of valuation that question the role of place in constructing adult education curriculum practice and how that constructs/produces difference. By asking the seemingly theoretical but essentially practical (and postmodern/poststructuralist) question of what does place have to do with identity and power, we enfold that question in what is nearly always taken to be a procedural question of no political import. Where do we "place" our adult education programs and what is the import of that placement? Our response shows how the construction of difference is tied to the production of power and identity in specific places. The point of the process of valuation is to examine relationally the production of difference through the production of power and identity in order to ask Harvey's (1992) question of how and why people invest places with power and how that power is used.

In what many take to be the apolitical, routine technical work of managing adult education curriculum development, an important but typically unasked question is how place contributes to producing the power and identity of participants. The social construction of place, and thus the production of identity as the interflow of power relations, can be analyzed in three ways: as the meaning attributed to specific physical places, as a community of social interactions in a specific place, and as a discourse about place, each of which contributes to producing power relations and identity. Earlier work[5] has used this analytical matrix to demonstrate how place, community, and discourse were relationally instrumental to producing/reproducing the identity of a group of professionals and the professional power they were seeking to achieve by locating their continuing education programs in particular places rather than others. Generally the study shows how place was crucial to the relational co-production of identity, power, and difference. More specifically, the findings show first how the physical location of a college campus contributed to constructing the professional identity and power of the participating professionals. Second, the findings show how the place of the campus helped to construct social networks associated with the prestige of academia rather than the privilege of a business resort. Finally, the study shows how the creation of professional discourse is shaped by its location in an academic place. Thus place, community, and discourse are used to comment on their role in producing/reproducing the identity this group of professionals was trying to construct for itself, as well as the professional power they were hoping to achieve through this process of locating the production of their professional identity in specific places. That identity and power combined to create the difference necessary to locate themselves in society as different than others and thus able to enact their power.

Place. Our assumptions about what place *means* are often left undisclosed and taken for granted. So what is the role of place in this continuing professional education program? In terms of seeing an academic location as a site for the enactment of certain social practices, the organizers of the continuing education, the participants in it, and the profession itself believed that locating their programs on college campuses represented important signifiers of identity and power and thus difference. The curriculum organizers[6] collectively presumed that by meeting physically in places called college campuses, they could then become deliberately associated with the putative authority and legitimacy of those academic enterprises. For the past 150 years any occupation that has successfully elevated itself in the US public's eye to the status of "profession" has done so by locating itself specifically in universities. These, in turn, have explicitly taken on the role of producing the scientific knowledge needed to identify the expertise of the profession as well as training the practitioners in the use of that expertise (Larson, 1979). This formula of linking knowledge production with professional training located in universities is so well established that most of us take it for granted. A number of leaders in this professional organization, however, did not. They deliberately worked to locate their programs on academic campuses because they believed such placement would be instrumental in constructing professional identity and professionalizing the occupation. So the importance of locating the program physically on a college campus had real as well as symbolic meaning. In deliberately associating the continuing education programs with the legitimacy bestowed by academic auspices, the association was signaling to its members and the public grounds for and power to do their work as "professionals" who had the requisite knowledge and training. Put another way, they were constructing their identity relationally by "being" in that place. Given that this occupation had historically not successfully managed their transition from trade to profession, the use of the "college" was quite significant in establishing its professional legitimacy, which depended, as Foucault would say, on establishing the knowledge/power couplet. Thus, a definer of difference is the "locating" on the college campus; such locating was key to establishing professional power.

Community. For much functionalist sociology and anthropology, community has consisted of the place in which face-to-face social interaction happens on a recurring basis (Giddens, 1990). While some might put forth the notion that such face-to-faceness has become both distanced and fragmented in our globalizing and telecommunicating world, for many humans the "compulsion of proximity" (Boden & Molotch, 1994) still remains an essential dimension to constituting human community. Boden and Molotch (1994) argue that community depends upon "the robust nature and enduring necessity of *traditional* [meaning face-to-face] human communication procedures." They argue that for biographical and historical reasons, "co-presence" remains "the fundamental mode of human intercourse and socialization" (ibid.). So, despite claims to the contrary, such "proximity" remains central to human community and, as Revill (1993) says, "has been championed as a source of identity, of moral and social stability, of shared meaning and mutual co-operation." This means it has an important "part to play in the way people think about themselves, in the construction of subjectivity" (ibid.). Revill (1993) further argues that the "social and physical boundaries and the rituals" that define communities "therefore become of paramount importance to the construction of community" because it is about "defining and ordering relationships between you and me, us and them." Revill suggests that communities shape the production of identity through informal social networks, the shared experience of people, and a sense of history.

Placing the professional program on the college campus was important for establishing the symbolic meaning of this continuing education program itself as well as distributing the material and symbolic benefits to its participants. That physical place was also important for creating space for co-presence. Of course, that co-presence would be possible at any site chosen for the program (or even mediated through various "distancing" technologies). But in this case the need to professionally legitimate the occupation by its location on a college campus was crucial to shaping the sense of community the curriculum organizers and participants were constructing. So it was important for that co-presence to be in a place that would define the meaning of the community in such a way that would promote a professional identity. Now it is fairly routine to hear professionals proclaim that while the organized events of any educational program or professional conference are important, the "real" reason they go is for the professional networking, the collegiality, and the "informal" interacting with others to make or continue important professional contacts. Thus the organizers' strategy as curriculum programmers was to provide a place in which those significant interactions could occur but would have specific meanings—deliberately relating professional learning with academic auspices—that would contribute to developing a sense of professional community. Although participants routinely challenged the locating of the program on college campuses, such locations were essential to the organizers in establishing the legitimacy of the profession in a way that a business or vacation resort (sites often requested by participants) would only speak to the privilege of established professions. By using the professional organization's and its participants' professed interests in continuing professional education, the organizers devised specific conditions that would also enhance the development of professional community by using the specific meanings associated with the college campus site. Further, creating these conditions for co-presence contributed directly to the development of a national community of professional identity as participants often came to a specific college campus from many different regions of the country. That identity helped establish difference, a difference defined by locating the program in a place whose authority legitimated the acquisition of professional knowledge and thus professional power.

Discourse. An important solidifier of place and community is discourse because it shapes "the logic of relations of contiguity in space and time and, in so doing, pattern[s] the organization and meaning of our lives" (Friedland & Boden ,1994). So how does discourse pattern meaning and life and in what ways is it associated with place? In his work Foucault (1970) repeatedly returned to the theme that different historical periods define different modalities of "truth." Modernity's truth, as he would say, is disciplinary. Disciplines represent productive relations of discourses and power (the knowledge/power couplet): "the exercise of power perpetually creates knowledge and, conversely, knowledge constantly induces effects of power" (Foucault 1980). Professional discourse represents dominant understandings of how the world works and people's relations to each other in it. Power emanates from these structures of knowledge that allows practitioners of disciplines ("professionals") to define others as "objects" of professional knowledge and manipulation (Wilson, 1999). The font of modern disciplinary power has increasingly resided over the last 150 years in academic places.

In this case, using plenary sessions at continuing education programs to introduce discipline theory for unifying the profession solidified the academic place and the construction of community. This was important as this occupation was comprised of several quite operationally different subspecialties, most of which recognized few or no discipline or practice connections with each other. Using Foucault, we want to suggest that the construction of a specific occupational discourse focused on producing a body of shared theoretical knowledge was essential to creating both a community of professionals and the knowledge/power they needed to identify themselves to the public as professionals. The relationship between place and community becomes more significant when tied into the creation of a specific disciplinary discourse itself located in and derived from a specific place. It now becomes clear the premium placed on associating this continuing education program with a college, why the place was significant in establishing the community whose professional power depended on defining disciplinary knowledge that too was defined in this college place. The difficulty facing this occupation in establishing its identity and power had been its historical failure to develop a disciplinary body of knowledge. As is the case with most professions, producing that disciplinary knowledge and training practitioners needed locating in an academic setting. Thus the continuing education programs were deliberately placed on an academic campus to constitute the professional community in a discursive fashion, which again was crucial to establishing how these professionals were constituting their identity differently than others in the same occupational group.

The Production of Justice

What do place, community, and discourse have to do with constructing/producing difference? The seemingly innocent, "technical" question of "siting" a program shows the political richness of understanding place as a producer of difference. Thus we argue that the locating or "placing" of our educational programs is a deeply political process and have used this case to show how place is implicated in the production of power and identity and thus difference. We have tried to show, using Harvey's process of valuation, how certain permanences get constructed and altered in the placing of educational curricula. In this case the permanences are small scale as we have looked at how place through the embodiment of location, community, and discourse are constitutive of power, identity, and difference in a specific occupation.

Following Harvey, we need to ask about the family of justice that these educational practices and consequences embody. It is difficult because contradictions now emerge that can seriously debilitate the work itself. Among possibly many, there are two dramatically different ways to assess justice in this situation. One includes the standards implicit in the professionalization movement itself. That view, briefly put, is that the events recounted here are indeed appropriate for justifying a differential deployment of power that enables some people to exercise that power over others, thus defining, through the use of discipline power, a difference of identity as well. It is just in other words that a class of individuals has the ability and authority to direct our lives by making choices for us through the acquisition of specialized knowledge and skill. Indeed, this is a very common way of defining the power of professionals. Would we want just anyone who felt like it tending to our health needs, building our buildings and airplanes, commanding our militaries, writing our laws, or doing any of those myriad activities for which the term "professional" now encloses? The program recounted here is thus just in facilitating the development of specialized professionals able to make important decisions for us, because we seemingly lack those skills and knowledge and thus are unable to make the appropriate decisions ourselves. A counter view, most visibly voiced by Illich and Foucault, would argue that such programs are deeply unjust because they create opportunities for selective individuals/institutions to acquire and exercise power over others. This acquisition and exercise of power is unjust because it disables humans (in Illich's view) and reduces them to objects acted upon by disciplinary powers (Foucault's view). The injustice results from people losing the ability to make real decisions about what happens in their lives. By using a process of valuation rather than a fixed notion of value we can begin to see a central dilemma of educational work even in the small scale recounted here. It does show how there are sides and how perforce through assumption, ignorance, or action we must choose. We believe that by asking about how permanences get constructed we can do as Harvey recommends, either reinforce or alter them according to our evolving notions of justice. If we don't invoke such interrogations, then it's clear that reigning systems of permanence remain intact as *someone* else's sense of what is just.

A GEOGRAPHY OF ADULT EDUCATION CURRICULUM PRACTICES

Where did all this production of identity, community, and discourse occur? One response is say that the place of the program mattered because the college campus itself was crucial to defining how these professionals were different from others practicing their occupation. As argued, the specific professional identity and invocation of professional power were possible only because of the location of the continuing education programs on college campuses. We have tried to use a critical questioning of the construction of place as a relatively enduring permanence to see how difference is produced as a way of asking about the justness of such production.

Adult education matters in the world because, as we have argued previously (Cervero & Wilson, 2001), it significantly contributes to the selective distribution of knowledge and power in society, as we think this chapter demonstrates. Key questions to ask when examining such distributions are who benefits from adult education and in what ways. When asked incisively, answers indicate that adult education not only inequitably distributes knowledge but the material and symbolic benefits of social, economic, and political power as well. Consequently, adult education is oftentimes a key producer of difference as its benefits in favoring some enable the production and exercise of power. That such distributions also disempower others is no secret. That's the point, that's why adult education matters. So we conclude by restating that adult education curriculum practice is better understood as a social process of negotiating interests in relations of power than as a procedural set of technical skills (Cervero & Wilson, 1994). Further, along with the instrumentally productive negotiations needed to construct actual educational programs (assessing needs, designing curriculum, managing sites, evaluating consequences), adult educators also through such negotiations socio-politically either maintain or change the working relations of power that enable them to exercise their instrumental activities (Cervero & Wilson, 1994, 1998). Thus, to understand the political consequences of everyday adult education practice, we have theoretically and empirically argued that adult educators in producing programs with specific outcomes either reproduce or change the relations of power in which they act. Part of understanding this simultaneous "acting in" setting and "acting on" the setting itself means understanding the geography of power represented in the working relations required to both produce programs as well as maintain or reconstruct the working power relations themselves. These working but changeable relations of power exist not just in time but also in space as located in and through particular places. In order to understand the production and/or transformation of power relations, we have to understand the role of place in producing power and how we as adult educators shape and are shaped by our "placeness."

In negotiating power to responsibly represent interests, we have also argued that adult educators are knowledge-power brokers who must make clear whose side they are on in a terrain constructed by the unequal distribution of symbolic and material benefits. In other words, there are no innocent or neutral spaces in the production of adult education curriculum. Adult educators' actions necessarily contribute to advantaging some to the exclusion of others. With a sense of how place plays in the production of power, a cartographical imagination (Edwards et al., 2002), we can better read, assess, and act in/on these terrains of power in order to ethically choose the right sides to represent in producing adult education and lifelong learning. Without it, we can't "see" the unequal relations of identity and power and can't imagine or construct new permanences that employ our visions of justice.

REFERENCES

Apple, M. (1996). *Cultural politics and education.* New York: Teachers College Press.

Bernstein, R. (1983). *Beyond objectivism and relativism.* Philadelphia: University of Pennsylvania Press.

Bird, J., Curtis, B., Putnam, T., Robertson, G., & Tickner, L. (Eds.). (1992). *Mapping the futures: Local cultures, global change.* London: Routledge.

Boden, D., & Molotch, H. (1994). The compulsion of proximity. In R. Friedland & D. Boden (Eds.), *NowHere: Space, time and modernity.* Berkeley: University of California Press.

Bondi, L. (1993). Locating identity politics. In M. Keith & S. Pile (Eds.), *Place and the politics of identity.* London: Routledge.

Cervero, R., & Wilson, A. (1994). *Planning responsibly for adult education.* San Francisco: Jossey-Bass.

Cervero, R., & Wilson, A. (1998). Working the planning table: The political practice of adult education. *Studies in Continuing Education, 20*(1), 5–22.

Cervero, R., Wilson, A., & Associates (Eds.) (2001). *Power in practice.* San Francisco: Jossey-Bass.

Edwards, R., & Usher, R. (2000). *Globalization and pedagogy: Space, place, and identity.* London: Routledge/Falmer.

Edwards, R., Cervero, R., Clarke, J., Morgan-Klein, B., Usher, R., & Wilson, A. (2002). Cartographic imaginations: Spatiality, adult education, and lifelong learning. In *Proceedings of the 43rd Adult Education Research Conference.* North Carolina State University, Raleigh, NC.

Foucault, M. (1970). *The order of things.* New York: Random House.

Foucault, M. (1980). *Power/Knowledge: Selected interviews and other writings, 1972–1977.* (Edited by C. Gordon). New York: Pantheon.

Foucault, M. (1984). *The foucault reader.* (Edited by P. Rabinow). New York: Pantheon.

Frieldand, R., & Boden, D. (Eds.) (1994). *NowHere: Space, time and modernity.* Berkeley: University of California Press.

Giddens, A. (1979). *Central problems in social theory.* Berkeley: University of California.

Giddens, A. (1984). *The constitution of society.* Berkeley: University of California.

Giddens, A. (1990). *The consequences of modernity.* Stanford: Stanford University Press.

Harvey, D. (1992). From space to place and back again: Reflections on the condition of postmodernity. In J. Bird et al. (Eds.), *Mapping the futures: Local cultures, global change.* London: Routledge.

Harvey, D. (1993). Class relations, social justice and the politics of difference. In M. Keith & S. Pile (Eds.), *Place and the politics of identity.* London: Routledge.

Harvey, D. (1996). *Justice, nature and the geography of difference.* Oxford: Blackwell.

Herod, A. (1997). From a geography of labor to a labor of geography. *Antipode,* 29(1), 1–31.

Isaac, J. (1987). *Power and marxist theory.* Ithaca, NY: Cornell Press.

Keith, M., & Pile, S. (Eds.). (1993). *Place and the politics of identity.* London: Routledge.

Larson, M. (1979). *The rise of professionalism.* Berkeley: University of California Press.

Lefebrve, H. (1974). *The production of space.* (trans. D. Nicholson-Smith). Oxford: Blackwell.

Revill, G. (1993). Reading rosehill: Community, identity and inner-city derby. In M. Keith & S. Pile (Eds.), *Place and the politics of identity.* London: Routledge.

Soja, E. (1989). *Postmodern geographies: The reassertion of space in critical social theory.* London: Verso.

Sork, T. (1996). Negotiating power and interests: A critical perspective. In R. Cervero & A. Wilson (Eds.), *What really matters in adult education program planning.* San Francisco: Jossey-Bass.

Wilson, A. L. (1999). Creating identities of dependency: Adult education as a knowledge-power regime. *International Journal of Lifelong Education,* 18(2), 85–93.

Wilson, A.L. (2000a). Place, power, and identity in continuing professional education. In K.P. King & T.R. Ferro (Eds.), *Proceedings of the second eastern regional adult education research conference.* University Park, PA: Pennsylvania State University.

Wilson, A.L. (2000b). Place matters: Shaping power and identity. *Proceedings of the forty-first adult education research conference.* Vancouver, BC: University of British Columbia.

Wilson, A.L. (2001). The politics of place: Producing power and identity in continuing education. In R.M. Cervero, A.L. Wilson, & Associates (Eds.), *Power in practice: The struggle for knowledge and power in society.* San Francisco: Jossey-Bass.

NOTES

1. "There are gritty realities out there, realities whose power often is grounded in structural relations that are not simply social constructions created by the meanings given by an observer" (Apple, 1996).
2. To our rather uncritical portrayal, we hasten to add that there are theoretical problems: "space" is neither a self-evident concept nor a transparent metaphor (Edwards & Usher, 2000; Friedland & Boden, 1994; Keith & Pile, 1993; Soja, 1989). For example, in the many forms of 20th century Marxist analysis, history is the presumed cauldron out of which springs the materialist conditions structuring society; space is "dead" in this view, whereas time is the active agent (Harvey, 1992; Soja, 1989). More recently in the proliferation of positional presentments, "space" has become a foundationalist metaphor for the conceptualization of identity politics (Keith & Pile, 1993). In either case the use of space has become problematic, one in its perceived non-effect and two in the presumed consensus of literal and metaphoric meanings. See Wilson 2000b.
3. See also Apple (1996) for his discussion of "structural relations that are not simply social constructions" and in particular Isaac (1987) for a closely related discussion of "realism," the properties of "enduring mechanisms" which structure experience. Harvey's analysis seems a practical corollary to Giddens' theory of structuration in which "structure is both medium and outcome of the reproduction of practices" (1979).

4. Harvey's invocation is a distillate of what Bernstein (1983) tagged the grand Either/Or. Either we establish Archimedean points of certainty from which all is appended or we are lost to the throes of relativism. The perceived way out of this dilemma is the nearly universal invocation of critique from critical social theorists over the past several decades. As Bernstein once remarked, though, at what point does critique collapse in upon itself, rendering us incapable of action? We think Harvey shows a possible way out by coupling critical processes (rather than foundational principles) with commitments to action.
5. See Wilson (2000a, 2000b, 2001) for a fuller accounting of the study reported below and its methodology.
6. Chapter co-author Wilson was the adult educator curriculum planner who worked with lay educators from the professional organization itself. The organizational leaders did not have any real awareness of their programs or organization as an adult education enterprise, although they were clearly focused on providing continuing education for the organization's members. Their rather common although naïve notion of continuing education was one of using established "experts" in their profession to teach the association members the "new" theory and technical skills deemed necessary to conduct professional practice. Being unreflective about and resistant to querying their ideas of education, the curriculum construction and pedagogical practices well represented what was tagged decades ago as the "update" model of continuing education, leaving little room for consideration of other models of continuing education such as reflective practice. Indeed, a market model of continuing education was a prime motivator for constructing a curriculum that could be "sold" . . . but that is another story.

CHAPTER 10

PICTURING PLACES IN THE ASSEMBLAGE OF FLEXIBILITY IN FURTHER EDUCATION

Julia Clarke

The picture in Figure 10.1 tells us about the places where Jane, cleaning supervisor at Shiptown College in South East England, spends her working day. When Jane starts work at five o'clock in the morning, "I'm the one who puts off the alarms and everything ... that's my office in the corner ... and attached to that is the store cupboard with all my cleaning materials. Every cleaner has to have a key, so I issue the keys out as they come in, and they sign in the book." This drawing was produced in the course of research into the ways in which notions of flexibility are "worked" or "played out" in the lives of people who work and study in two further education colleges in Southern England. Embedded in a history of Craft Guilds, Mechanics Institutes and Technical Colleges, the boundaries of Further Education (FE) have been stretched since the early 1990s to incorporate adult, community and workplace learning in diverse locations. In England, FE colleges have now been subsumed within the Learning and Skills Sector, which encompasses all non-university vocational training and adult continuing education. By the end of the 1990s, further education colleges were a central focus for the UK government's agenda of widening participation in lifelong learning. National and international policy documents for post-school education and training in the 1990s abounded with

Space, Curriculum, and Learning, pages 139–155

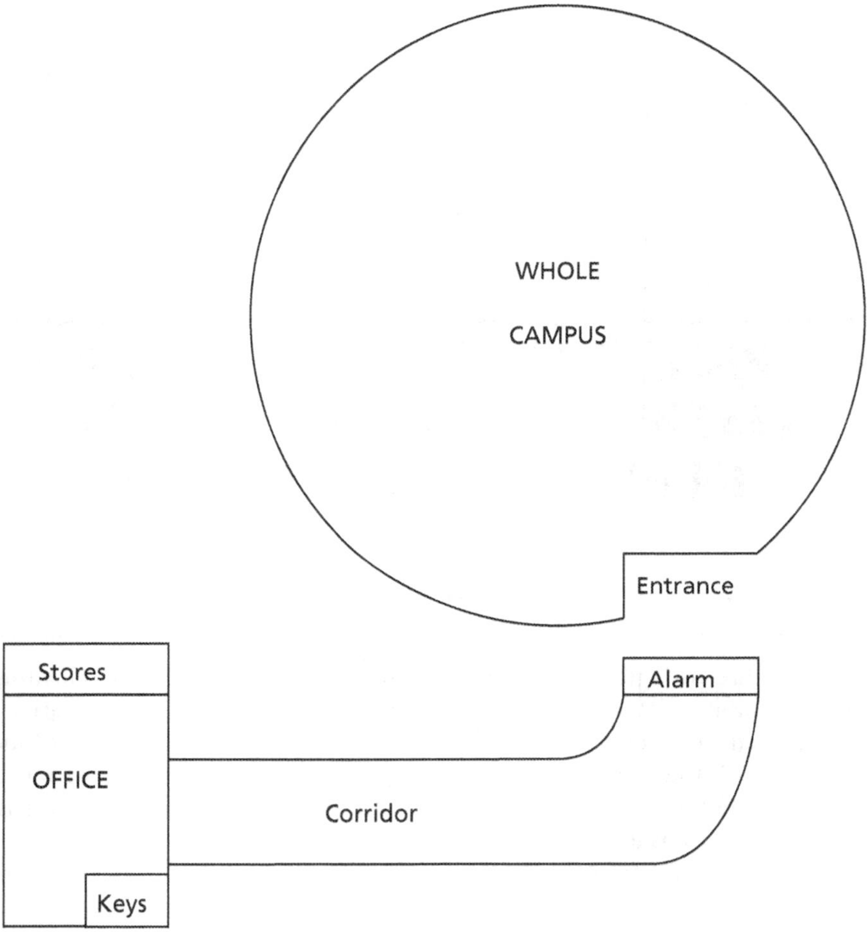

Figure 10.1. Jane's spatial map

calls for flexible forms of learning and assessment, for the development of flexible skills for work and flexible responses to rapid and unpredictable change (OECD, 1997; DfEE, 1998; Edwards et al., 1999).

In the course of my interview with Jane, she told me that she had "seen a lot of changes" during the twenty-seven years in which she had been cleaning the classrooms and corridors, workshops, studios and kitchens at Shiptown College. Flexibility, for Jane, was experienced as the effects of changing contractual arrangements: "I am here at quarter to five in the morning opening up. It used to be six when I first started, but because we can't get the cleaners now, we are having to start earlier, and the ones that I have got, and they do more hours. That is where the flexibility comes into it. Now, it is very, very flexi." Ten years ago, there were ninety-six cleaners, but "the new contractors came in and we lost half of that force, because ... the wage dropped immensely ... it really did drop. Anyhow, we lost half our staff; I stayed on for the simple reason it was near to where I lived, I was in walking distance ... so I didn't have to include bus fares. I wasn't happy with it, mind you, wasn't happy with it." Jane told me that, when she first started working at the college "...they had loads of caretakers.... Caretakers did the toilets, stairways, and corridors, empty the rubbish. That all came to the cleaners." Traditionally, caretakers were men and cleaners were women. So at a time when sociologists were writing about the "housewifization" of labour (Mies, 1986; Hart, 1992), the effect of these changes for Jane and her colleagues was the addition of "men's work" to their daily routine. In our conversation about her work in the college, Jane talked about her desire to keep the college "looking nice," about the need to comply with Health and Safety regulations and have the floors washed before students come. She said that one way to draw the attention of senior managers to the pressures on her workforce would be to withdraw the services of the office cleaner who "... does special things, like wash their cups up."

When I asked Jane to draw a diagram of the places where she works in the college, she was reluctant at first "Oh, I am not any good at drawing! No. You're asking me to do something there Julie." In my subsequent prompting, I said, "...perhaps just put in the middle where you spend most of your time, or where you think the most important place is." Here we can see how space is not just the scenery for things that happen, but is constituted in social relations of space and time. Jane said that she spent a lot of time in her office, but ignored my suggestion that she could represent this in the middle of the page. After drawing the office, stores and key cupboard in the bottom left hand corner, Jane launched into a chronological account of her working day. This began at the college entrance, which she drew with the alarm on the right hand side of the page. The corridor was then drawn from right to left, almost, but not quite, connecting two different temporal articulations of her working day. On the left are places where Jane spends "a lot of time," and on the right are the places that mark the beginning of her working day. In spite of my interventions in this process, the most striking feature in Jane's picture is the circle, labelled "whole campus," which represents a challenge to my assumption that some places were more important for her than others. When she began to question this assumption, I said, " ...just put a big box saying 'whole college' or something...." I guess I might have been anxious about getting through the interview schedule in time for my next appointment. But now Jane had lost her reticence and she drew the circle with a flourish, enclosing all the places she had been telling me about. Without embarking on a discussion of the iconography of squares and circles at this point, I suggest that this drawing reproduces a moment of tension between Jane and me. I wanted to dissect Jane's experience of a place in the interests of research, while her resistance to this is expressed in the bold encircling of the "whole campus."

In this chapter, I want to explore these connections between time and place, visual and verbal representation, work and learning, and the processes and products of research. The title of this chapter carries two propositions. First, you are invited to join me in the process of picturing places through the words and drawings of just a few of the fifty staff and students who were interviewed in the course of an empirical study in two colleges—Shiptown and Techcity. This will be a largely descriptive account, with the addition of interpretative questions and suggestions arising from diverse interests to suggest other possible lines of inquiry. The second proposition in the title is that flexibility in further education is a particular kind of assemblage, but I should warn you now that I have no intention of reaching any conclusions about what kind of a beast this is. The kind of knowledge that generates such certainties is dependant on a binary logic that Deleuze and Guattari (1988) characterize as the tree model of knowledge, with its logical hierarchy of root, trunk and branch.

THE ASSEMBLAGE AS A SYMBIOTIC RELATIONSHIP

The places that feature in this chapter have been "pictured" in the course of a research project about which a number of stories can be, and indeed have been, told (Edwards et al., 2000; Clarke et al., 2002). Flexibility features in descriptions of the changing workplaces of global capitalism; of organizational structures and management practices; of the processes of teaching and learning; and in the construction of learners and workers as enterprising individuals with flexible skills and dispositions. This literature provided the conceptual framework for a study that was formulated by Richard Edwards in 1999, starting from a critical approach to notions of "quality assurance," "the multi-skilled worker," "the enterprising self" the "learning organization" and similar evocations of a "new work order" for which there has been little empirical evidence (Du Gay,1996; Gee et al., 1996). I was employed to pilot a methodology for this research, and together with two colleagues, Roger Harrison and Fiona Reeve, conducted interviews with fifty students, lecturers, managers and support staff in two colleges. Each interview lasted about an hour and was taped and transcribed, using fictitious names chosen by the interviewees. At meetings held less than a year after our first pilot interviews, we fed back some of our stories, along with a more conventional thematic analysis, to those who had participated in the interviews. There we presented one interviewee's understanding that flexibility was simply a substitute for "inclusion," the notion that had driven the previous year's policy initiatives, funding regimes and strategic plans. A senior manager at the meeting commented that flexibility was no longer the buzzword in the college, and the latest new idea was "leadership." If there ever was an assemblage of flexibility in further education, this suggests a particular expression of interests and desires taking flight from one assemblage and attaching itself to another.

So why bother with this assemblage if it has already metamorphized into something else? The answer is precisely because we should be suspicious of the constant re-branding of policies and strategies with new "buzz words" that determine what kinds of knowledge will be valued as current, relevant and useful to the enterprise of lifelong learning in a global economy. Deleuze and Guattari's analysis of this phenomenon of "unlimited semiosis" in the relationship between capitalism and schizophrenia is far too complex and wide-ranging to try and summarise here. However, it is important to know that the notion of the assemblage is grounded in a link between the Freudian concept of libido and the Marxist concept of labour power. The argument is that, while capitalism created divisions between politics and desire, psychoanalysis reproduces the split between the individual and society (Holland, 1999). The assemblage reaches across these divisions in a symbiotic relationship between content and expression, in which "content is not a signified, nor expression a signifier; rather, both are variables of the assemblage" (Deleuze & Guattari, 1988). These variables comprise, on the one hand, the "machinic assemblage of desire" in whom bodies, actions and passions are mingled in particular relationships in the material world. On the other hand, the "collective assemblage of enunciation" refers to the words, acts, statements that transform the corporeal world in a particular "regime of signs." The term machine is used to denote an interest in how an assemblage works and what it does, rather than in what it is or what it means.

The assemblage is one of many bewildering, tantalizing and impossibly difficult concepts produced in Deleuze and Guattari's monumental "thought experiment", which has to be understood in terms of their rhizomatic qualities (1986, 1988, 1994). As a movement in thought and politics that is both flexible and nomadic, Deleuze and Guattari offer the rhizome, with its subterranean, multi-directional connections, as an alternative to the tree model of knowledge. The qualities of the rhizome are those of connection, multiplicity and the ceaseless movement of subterranean stems establishing connections between disparate signs, substances, entities or modes of thought. Embracing these qualities, however, doesn't mean doing away with the products of structural thinking, but allows for different modes of thought to connect with rhizomes and take off along their multiple threads.

"To be rhizomorphous is to produce stems and filaments that seem to be roots, or better yet connect with them by penetrating the trunk, but put them to strange new uses" (Deleuze & Guattari, 1988).

A tree-like account of our research on flexibility in further education would begin at the roots with a rationale. The branches of theory and method would grow from the solid trunk of "the literature," to feed the leaves, flowers and fruit of analysis, interpretation and conclusions. I began instead with Jane's drawing because it felt right to bring to the fore both her experience of the changing relations of capital and labour and her concern with the physical spaces and places of the further education college. Jane's drawing is neither an expression of her subjectivity, nor is it the college, nor a representation of the college. The drawing is all of these. It is also a map of bodies and desires, of signs and symbols, objects and social relations, of material places and of many possible readings and representations of these. While the tree model generates tracings that merely reproduce an image of a known world, the map, in Deleuze and Guattari's view, is a performance that is "... open and connectable in all of its dimensions; it is detachable, reversible, susceptible to constant modification" (1988). And the map is one layer, or stratum in the assemblage that I have posited in this chapter as the assemblage of flexibility in further education.

WORDS AND PICTURES: ANOTHER WAY OF TELLING

Among the claims regarding the use of visual data in social research, there is an idea that a map or drawing can help us to know something "...so we can 'see' something different" (Paulston, 2000). Visual anthropology is characterized as "... a field of social action which is enacted in planes of time and space through objects and bodies, landscapes and emotions, as well as thought" (Banks, 1998). Fischman argues for "...the incorporation of visual culture into educational research" in order to investigate a territory "... that holds layers of meaning we may not otherwise have encountered (Fischman, 2001). And a photograph can evoke visual memories that are "...freer than reason" (Berger & Mohr, 1982).

The decision to ask for drawings in the course of our research interviews was made in the spirit of open-ended experimentation rather than as a clearly developed strategy. Most of the participants were happy to talk, knowing their voices were being recorded, and that these recordings would be transcribed into written texts. They were generally quick to respond to questions, without asking for clarification or explanations, and there were only very occasional expressions of uncertainty about the choice of words in the formulation of their answers. It was very different when we asked people to draw a picture, map or diagram, and this is partly a reflection of our own ambivalence about making such a request. Instead of equipping ourselves with a careful selection of drawing materials, we generally presented people with a page torn from a notebook, and whatever kind of pen or pencil came to hand. When there was a shortage of time, this was usually the section that got left out, and the transcripts show signs of hedging and hesitation when we made the request. Responses are often prefaced with self-deprecating remarks, "I'm no good at drawing" or "I'm not an artist," and the drawings were rarely produced without prompting from the interviewer with regard to both content and forms of representation. While I would not claim that the drawings presented in these pages offer any kind of privileged access to the subjective experience of the people who produced them, I do want to argue that different forms of representation produce different kinds of knowledge. It is important to emphasize the extent to which each of the drawings presented in the following section is contingent upon a whole series of interests, desires, feelings, interactions and events colliding at the particular moment when they were produced.

PICTURING PLACES

George, a student at Techcity College had a heavy cold on the day of our interview. This may account for the prominence of bedroom and bed in his drawing (Figure 10.2), although George did say that he studied at home and "at the moment most of my books are at the bottom of my bed…" The college is several miles away from George's home, but the absence of a fourth wall to his bedroom would allow a body to float between his desk at home and the desks and computers in the college library. As a mature student, George found it difficult to study in the library because the younger students are too noisy: "they think that the library's some kind of meeting place… what I do is try to find a nice quiet corner. I mean it's just convenient in terms of having the computers as well, books, there..."

I asked George why he didn't include any classrooms in his drawing, and he said that, although "lectures actually happen in classrooms" he needed to "…re-digest most of the information I get anyway for it to kind of start to be absorbed as it were... at another point." The relationship between places in George's drawing are structured by an orientation towards individual study, and much of our conversation was about fitting this into the rest of his life. He was hoping to go to university after a year on a full-time access course, because he felt that "…having a degree gives you a lot more freedom and flexibility in terms of the kind of work you do."

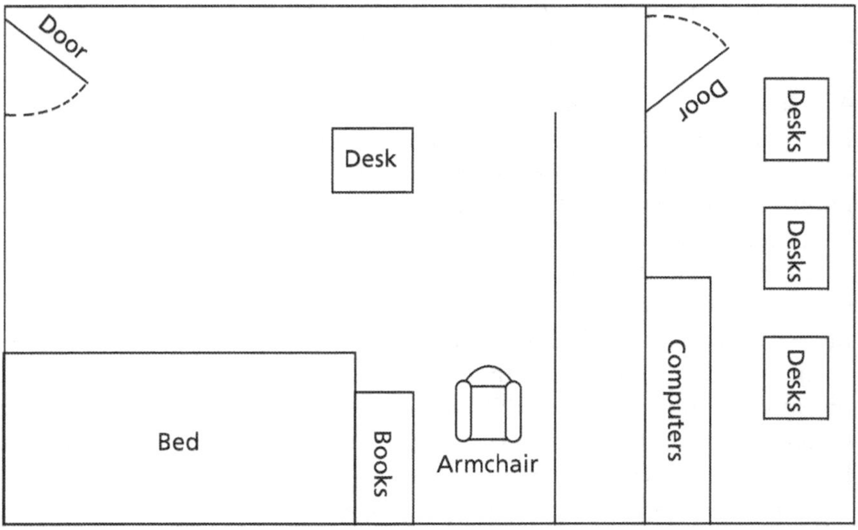

Figure 10.2. George's spatial map

For John Perry, a young student at Shiptown College, on the other hand, the college campus appears to be the location of social interaction (Figure 10.3). The first place John drew was the bike shed, and he talked about locks and security cameras and his fear of his bike being stolen. John was following a course that will qualify him to join the police force. Throughout the rest of the conversation that accompanied the drawing, John Perry used the collective "we," identifying one of the classrooms as "our base room" and including both the canteen and L Street, which he explained as "...just the building over there, it's like lunch time... get our sandwiches or play pool."

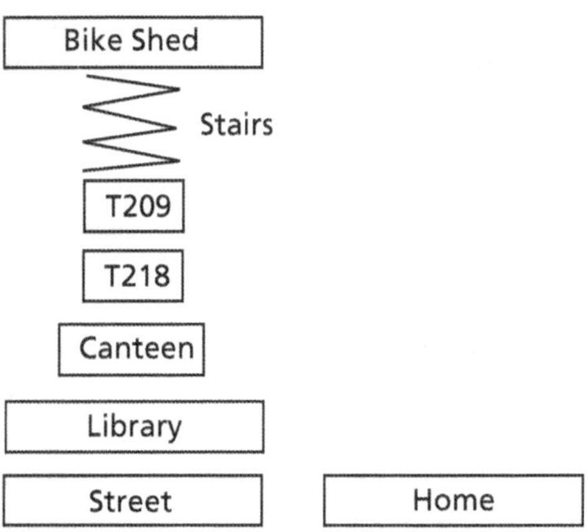

Figure 10.3. John Perry's spatial map

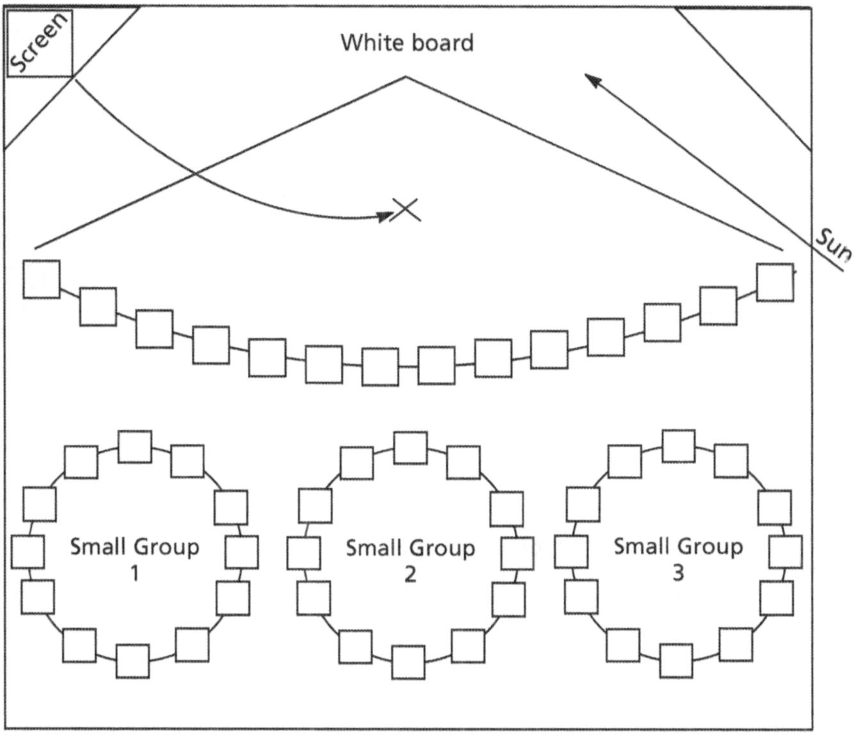

Figure 10.4. Legs' spatial map

Legs is a part-time tutor at Shiptown College, whose work in the college is represented as the classroom (Figure 10.4). As a deaf teacher of British Sign Language, Legs was preoccupied with the importance of visibility. The drawing shows us how he arranges the seating for large and small groups, and indicates the problem of the sun reflecting on the white-board. The drawing also shows how a solution to the problem would be to have a screen on a trolley that could be moved to the middle of the room.

To watch a fluent sign language communicator describing a place is to experience being in that place in a way that is qualitatively very different from hearing or reading a verbal description. The signs and gestures locate people and objects visually in a physical and affective relationship with the narrator. In Legs' drawing, there is a far stronger sense of the relationship and movement of objects in space than in any of the other drawings. Perhaps this is why a single classroom is all that we see.

There is no reference to classrooms in Beth's drawing (Figure 10.5) of the places where she works as a senior manager in the same college as Legs, John Perry and Jane. Beth started her drawing with the square at the top of the picture, and as she drew she explained,

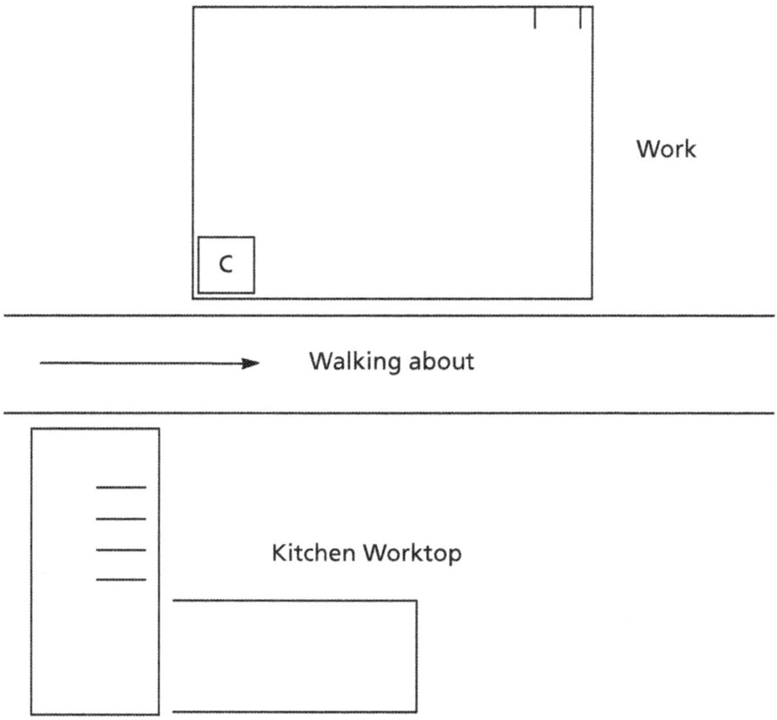

Figure 10.5. Beth's spatial map

> "… there is the door and the key bit there is the computer. So, that is work… I suppose, as well if I did the corridor, because I think it is walking about. You can do an awful lot of really good work by …managing by walking about, and that is important. And… when I go home, the bit for me is the kitchen. And, it's a kitchen worktop really, so that the stove is there, so dinner's cooking, but then there is papers all on here, because I am working along the kitchen worktop… the other bit is the laptop, which gets involved in that as well."

This particular combination of drawing and talk brings us into Beth's habitual presence, in which she pictures herself at the computer "managing by walking about" and "working along the kitchen worktop." Beth furnishes her kitchen with stove, worktop and papers, while "dinner's cooking" and the laptop "gets involved." Time is certainly implied in the movement between places and in the simultaneous activity of cooking and working with papers, which invites us to pose questions about the relations of time, space and gender. Is it a common practice for senior managers to deal with paperwork on a surface designed for chopping onions, while their dinner is cooking? Do men and women produce different kinds of spaces, and different relationships between the spaces at home and at work? Or would differences of social class be more marked than those of gender, for example in the ways that Beth and Jane each inhabit and produce these spaces? And why, or for whom, do these differences matter?

The central location of corridors in both Jane's and Beth's diagrams suggest that these are places to which strong feelings are attached, even if the meanings are very different. For Jane, the corridor was a route, connecting her office to the rest of the campus, but when Beth talked about her work, she said that "walking the corridors is absolutely vital... it is very much moving about, and going round and seeing." In her study of "parental discourses of control" in UK further education colleges, Leathwood (2000) cites a woman lecturer in a further education college complaining about "corridor meetings where decisions are made' as symptomatic of 'a very male way of doing things." Unless we characterize Beth's behaviour as male, then her practice of "managing by walking about" requires a different interpretation. Beth's male colleague, Barney, chose to write a list when I asked him to produce a diagram. But he included corridors in his list, saying that, "I make an effort to go round and see all the staff in the morning, talk to them, see if there are any problems. Any minor issues that arise here rather than ringing people I tend to walk and see them.... Ok it's obviously not private and confidential but if it's something you need to see a member of staff on, or need to have a quick word, you work in the corridors."

When they walk the corridors, Barney and Beth may be demonstrating caring and responsive styles of management while they are simultaneously engaged in acts of policing or surveillance. In "Discipline and Punish", Foucault (1977) describes the way in which docile bodies are induced to discipline themselves through practices of visuality in which they take on the monitoring, scrutiny and appraisal of their own practices as if they are constantly subject to the gaze of unseen others. While corridors might thus be construed as part of the disciplinary architecture of educational institutions, the tower in Christy's diagram (Figure 10.6) bears a stark visual resemblance to Jeremy Bentham's panoptican, which Foucault takes as a material expression of this notion of surveillance.

Bentham's plan for a panoptican included a tall tower encircled by buildings in which the windows, blinds and corridors were arranged so that the occupants of individual cells could be observed from the tower. There is a striking similarity between Bentham's plan and Christy's tower, with the dotted lines fanned out from its base in a circular direction. Christy is a senior manager in Techcity College, but when he drew the tower in Figure 10.6 he seemed to be more preoccupied with how he was seen by others than with his view of them: "I'm sure this is how people would see us. The senior management team is on the sixth floor, the director's floor of the college, so that's physically where I work."

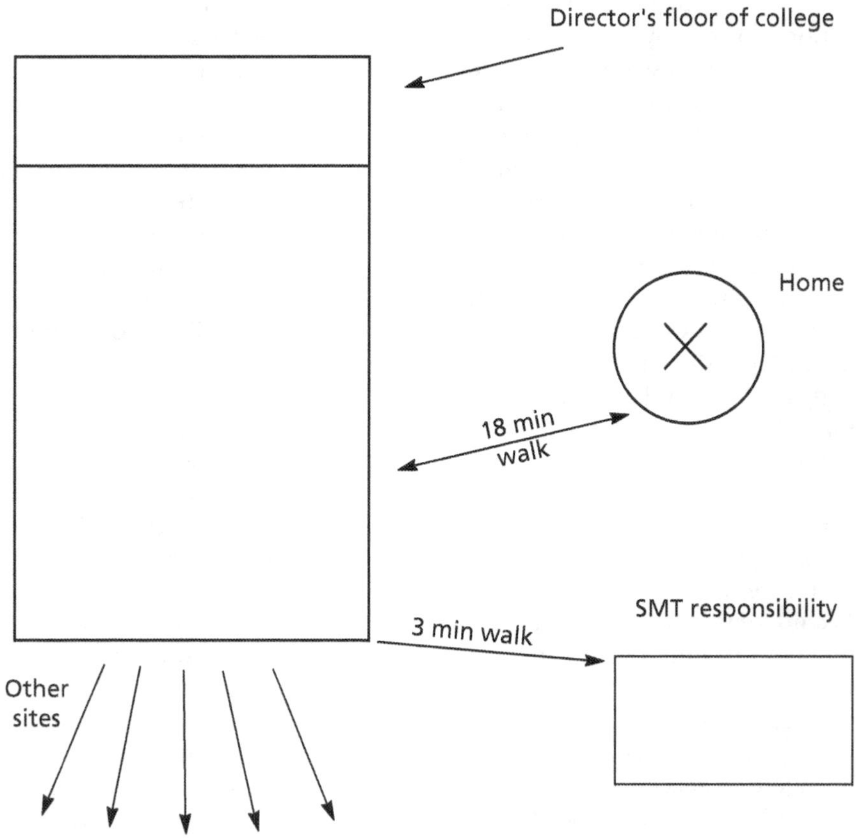

Figure 10.6. Christy's spatial map

Earlier in our conversation, we were talking about a plan for the college to offer a more flexible programme that included opening at weekends. Christy saw this idea as part of a "vision" of flexibility that he wanted his staff to embrace: "I think as a service to the community it's crazy that we close at weekends." He added that changing this would mean weekend working for senior management since "we should be with them, we should be seen to be around." But the strong precision of the arrows between home and work point to a conflict between this assertion of responsibility and a more compelling desire. "It's an eighteen minute walk so when I leave here I know I will be home, bar being run over by a car, in eighteen minutes maximum and that's a great certainty actually in my life." Does this evocation of a desire for certainty in his life explain the emphatic repetition of the phrase "walking the job" in Christy's retraction of his earlier statement about the need for senior managers to be "seen to be around?"

"I don't concur with people who say you've got to be constantly be seen walking the job. I don't think it makes much difference to anybody's working day whether they see me walking past the office or classroom. My job is to steer the strategic direction of this college... but walking the job, I'm not necessarily into that walking the job."

In these extracts from the transcripts of our conversation, I have represented aspects of Christy's experience that are not expressed in his diagram. On the other hand, there are meanings to be discerned from the visual image that weren't expressed in words. There is an obvious correspondence between the rectangles in the drawing and the physical shapes of the buildings to which they refer, but the representation of "home" by a circle with a cross inside it is clearly symbolic. Kress and van Leeuwen (1996) sum up the meanings attached to circles and squares in contemporary Western art and design in terms of a binary opposition between a natural and a technological world. In this analysis, curves and circles are associated with qualities of warmth, protection and timelessness "... and such mystical meanings as may be associated with them derive from this." Straight lines, squares and angularity, on the other hand, are associated "... with the world of technology which is a world we have made ourselves, and therefore a world we can at least in principle understand fully and rationally" (Kress & van Leeuwen, 1996). When this kind of binary is expressed in words, we can deconstruct it by looking for ways in which, for example, nature is repressed in the articulation of social production or the social is repressed in the representation of nature. In these visual images, however, the tensions between straight lines and circles suggest a different relationship, in which ambivalence and ambiguities are brought to the surface in the same space. Rose (2001) points out that iconography depends on intertextuality for its interpretative power, and drawing on other visual texts, I associate the cross inside Christy's encircled "home" with a special place on a map or a photograph, the site of hidden treasure or the location of a friend's house. But if Christy's drawing tells us, "this is where you'll find me," then the fluid lines of flexibility are bound to disrupt the arrow of certainty, running on its precisely timed trajectory from home to work and back home again.

Audrey also expressed ambivalence in our interview conversation, but here the tension is not between home and work, but between her work as a lecturer at Techcity College and the bureaucratic and managerial functions associated with Christy's tower. This is expressed in Audrey's combination of material and symbolic representations, in which the tower features as an "incredibly important place" that "... makes my life a misery at times, or it makes my life nice at times." Audrey is a lecturer at Techcity College, based at one of the "other sites" represented at the end of Christy's dotted lines. Audrey began this drawing (Figure 10.7) with the cloud-shaped classroom in the centre of the page, and then proceeded to add places, objects, people and events in constellations around this focal point. Although Christy said that he occasionally took work home, in both his talk and his drawing, home was kept entirely separate from his work, and it was this attempt to separate the two that appeared to be problematic. Audrey, on the other hand, spoke about home as the place where she does her preparation for work, saying that this was as important as the classroom. Hence the similarity between the shape, size and central location of classroom and home in her drawing. She drew lines to link bed, desk and computer with home, and then Audrey said that she thinks about work on the tube, but corrected this to "train" and added "walking home" as well. The evident sites of tension in Audrey's drawing are the two towers, standing in stark contrast to the images of movement and connection that they partially enclose.

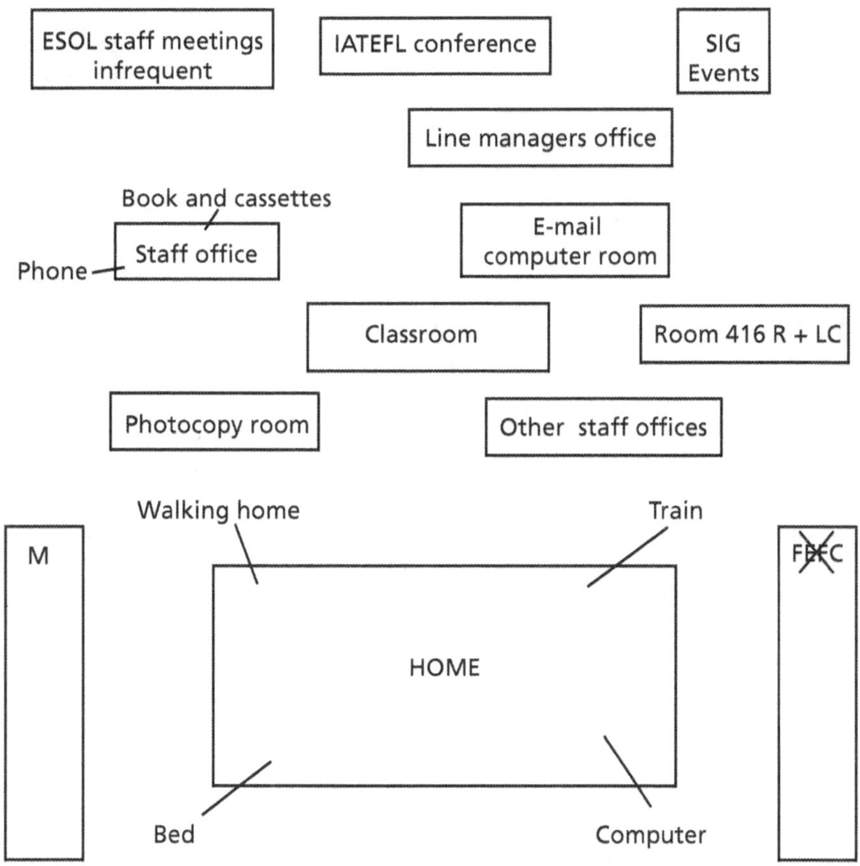

Figure 10.7. Audrey's spatial map

Audrey referred to the tower, M, on the left as the Headquarters, commenting on its importance as a source of both misery and something "nice." She then added the matching shape on the right hand side and labelled it FEFC, signifying the Further Education Funding Council. This was the national funding and regulatory body for further education at the time, (now replaced by regional Learning and Skills Councils). I asked Audrey whether the FEFC was also housed in a tall building, and she responded with an apology for "getting carried away" and crossed out the initials, saying "... FEFC isn't, but M well it's there and I don't go to it very much but it has an enormous significance on my work." Thus Audrey's drawing establishes a symbolic resemblance between the material structure of "the Headquarters," and the imagined place of government offices, both represented as the hard and static structures of control, contrasting with, and setting partial boundaries around, the soft and fluid structures of classroom and home.

BODIES, KNOWLEDGE AND PICTURING PLACES

So what have we learnt from these drawings about the notions and effects of flexibility in the lives of people like Audrey, Christy, Beth or John Perry? Nothing and everything. We've learnt nothing that we can be certain about, since certainty depends on eliminating variables in order to observe uni-directional relationships of cause and effect. The sense in which these images can tell us everything is that they include a multiplicity of variables that are related to each other in the assemblage of flexibility in further education. This is not the assemblage of "inclusion" or "leadership," or whatever next year's buzzword happens to be. Nor is it simply the assemblage of further education. The encounters that generated these drawings were all framed by a particular interest in four domains of lived experience that are brought together in the discourses of flexibility. First, having defined the wider context in terms of the economy and labour markets, we asked questions about personal experiences of work, perceptions of change in working lives and the relationship between work and lifelong learning. Second, our questions about the college organization focused on flexibility in relation to place, time, and funding and institutional change. This was the point at which we asked interviewees to draw the map or diagram of places they associated with work or study in their college role. Our third set of questions was about the processes of flexible learning. Here we asked about experiences and perceptions of different modes of learning, including the uses of technology. Finally, we asked how people felt about their particular responsibilities, how they saw themselves in relation to their descriptions of other staff and students in the college, and how they felt about the articulation of these relationships in terms of providers and customers.

The forms of knowledge we produce within educational research both shape and support particular regimes of practice and define the objects—places, bodies, teaching learning, and work—of these practices. What we know always depends on how we come to know. Spatiality was not the focus of our enquiry but it cropped up everywhere. The first set of questions elicited stories of movements and migrations. Responses to the questions about teaching and learning were littered with spatial metaphors, and the descriptions of other people in the college included comments on physical characteristics, clothes, and bodies that are "... understood and lived spatially as much as are topographical sites in the landscape" (Shields, 1997). The drawings, despite the reticence with which they were produced, provide us with access to variables in this assemblage that can't be reached through words alone. Jane's "whole campus," George's bed, John Perry's bike shed and the sun shining through Legs' classroom window are all variables in the assemblage of flexibility in further education. So, too, are Beth's kitchen worktop and the corridors where she "manages by walking about," and Christy's 18-minute walk between work and home. The relationship between the tower in Christy's drawing and the double towers in Audrey's, is mediated by other relationships between enunciation (Christy's arrows, Audrey's clouds) and desire (the segmentation or immanence of work/home) enacted in each of their drawings.

Thinking about space means grounding our perspectives in our bodies and connecting knowledge with desire. In the symbiotic relationship of content and expression, "desire is, to some degree, the complete force field of all events, on all sites, encompassing all potentialities" (Howard, 1998). The potentiality of the assemblage is one that enables us to stop asking "what is this?" or "what does this mean?" and to focus instead on "how does this work?" and "what does this do?" When we take things apart to look at the pieces, picturing places is just one more way of putting them back together again.

REFERENCES

Banks, M. (1998). Visual anthropology: Image, object and interpretation. In J. Prosser (Ed.), *Image-based research: A sourcebook for qualitative researchers*. London: Falmer Press.

Berger, J., & Mohr, J. (1982). *Another way of telling*. New York: Pantheon Books.

Clarke, J., Edwards, R., Harrison, R., & Reeve, F. (2002). Why don't you give the money back? questions of accounting and accountability in three accounts of educational research. *International Journal for Qualitative Studies in Education, 15*(5), 555–569.

Deleuze, G., & Guattari, F. (1986). *Kafka: Toward a minor literature* (trans. D.Polan). Minnesota, University of Minnesota Press.

Deleuze, G., & Guattari, F. (1988). *A thousand plateaus* (trans. B. Massumi). London: Athlone Press.

Deleuze, G., & Guattari, F. (1994). *What is philosophy?* (trans. G. Burchell and H. Tomlinson). London: Verso.

DfEE (Department for Education and Employment). (1998). *The learning age: A renaissance for a New Britain.* London: The Stationery Office Ltd..

Du Gay, P. (1996). *Consumption and identity at work.* London: Sage.

Edwards, R., Nicoll, K., & Tait, A. (1999). Migrating metaphors: The globalization of flexibility in policy. *Journal of Education Policy, 14*(6), 619 –630.

Edwards, R., Clarke, J., Harrison, R., & Reeve, F. (2000). *Flexibility at work: A study of educational institutions.* Colloquium papers presented at the Working Knowledge conference. Sydney, December.

Fischman, G. (2001). Reflections about images, visual culture, and educational research. *Educational Researcher,* November.

Foucault, M. (1977). *Discipline and punish: The birth of the prison* (trans. A. Sheridan). London: Allen Lane.

Gee, J. P., Hull, G., & Lankshear, C. (1996). *The new work order.* Sydney: Unwin.

Hart, M. (1992). *Working and educating for life: Feminist and international perspectives on adult education.* London and New York: Routledge.

Holland, E. (1999). *Deleuze and Guattari's anti-oedipus: Introduction to schizoanalysis.* London and New York: Routledge.

Howard, J. (1998). Subjectivity and space. In E, Kaufman & K. Heller (Eds.), *Deleuze and guattari: New mappings in politics, philosophy and culture.* Minneapolis: University of Minnesota Press.

Kress, G., & van Leeuwen, T. (1996). *Reading images: The grammar of visual design.* London: Routledge.

Leathwood, C. (2000). Happy families? Pedagogy, management and parental discourses of control in the corporatized further education college. *Journal of Further and Higher Education, 24*(2), 163–182.

Mies, M. (1986). *Patriarchy and accumulation on a large scale.* London: Zed Books.

OECD (Organisation for Economic and Co-operative Development). (1997). *Sustainable flexibility: A prospective study on work, family and society in the information age.* Paris: OECD.

Paulston, R. (2000). *Social cartography: Mapping ways of seeing social and educational change.* New York: Garland Publishing.

Rose, G. (2001). *Visual methodologies: An introduction to the interpretation of visual materials.* London: Sage.

Shields, R. (1997). Spatial stress and resistance: Social meanings of spatialization. In G.Benko & U. Strohmayer (Eds.), *Space and social theory: Interpreting modernity and postmodernity.* London: Blackwell.

CHAPTER 11

DIMINISHING SPACES FOR LEARNING

The Supervision of Research Degree Students in Australian Universities

Pam Green

In a context where timely completions of research theses have become paramount and where the nature of research degrees is shifting dramatically, the space reserved for research supervision has come into the limelight. With a greater focus on supervision, as evidenced by the emphasis on supervisor training and policy development in Australian universities, there emerges the issue of the spaces remaining for, or devoted to, supervision. This chapter will consider notions of spatiality, curriculum and learning with respect to research supervision in the context of the changing research environment brought about by the emerging research performance economy in Australia. The diminishing space deemed permissible for research degree completions is likened to a staged performance in which supervisors are positioned as directors and where backstage preparation or supervisory practices are constrained by time and place. The first part of the chapter deals with the changing higher education context in Australia and the consequential rationed space for supervision on (or behind) the research stage.

Space, Curriculum, and Learning, pages 157–171

The relational, intellectual and physical spaces (RIP) for supervision in times of performance-driven approaches to university research education provide the focus of the second part of the chapter. In such times of diminishing space the questions that confront us include: what spaces are afforded to supervision? What supervisory practices thus emerge? What space remains for learning in the context of research supervision? This chapter explores these questions and the implied challenges for the future in terms of the "new space" for supervision in Australian universities. It is argued that the space is precious and pressured. Indeed the remaining space for supervision may not be a place at all in terms of supervision as we have known it. There appears then to be no place for space? Or is there?

Paradoxically we are left to make the most of the new stage on which we (or rather our students) must dance, even though the pace may not fit our view of what should lead to, form the basis of, and constitute performance. The development of a repertoire of completion strategies and a supervisor kit bag of management secrets might seem mandatory in the race to complete. However, we must resist the push towards reductionist approaches to supervision and argue for the acknowledgement of the complexities inherent in supervision. Paradoxes abound. If we dare to continue to treat supervision as a social practice and work with our students holistically, viewing them as people with complex and busy lives as well as knowledge workers creating a thesis or other form of durable record, how do we deal with the pressure to choreograph timely performances? If we view the expansion into new knowledge as a reflective, sophisticated process in which new ideas are formed slowly, tested and developed, how do we assist candidates in the rush to create unique and significant knowledge? How do we keep students dancing when the tempo is so much faster and the time backstage is so shortened?

These questions may well be viewed as a rejection of the need for change. This is not the case. Indeed change is all around us. In academic life, we are reacting to the change from a research economy located in traditional academic paradigms to one located in the context of performativity. Usher's (2002) "all change" metaphor, derived from childhood train journeys, shows the impact of the shifting context. While some within the academic community have heard the cry to change trains and moved platforms and consequently embarked on new journeys, some have moved to the wrong platform, and others have not heard the call at all. Therefore what matters is how we respond to change. We need to reflect upon, and act on, changes in our midst within higher education so that we are not left standing but also so that we can move onto meaningful journeys rather than be pushed blindly into directions unknown.

Research and research training arenas in Australia are now subject to ongoing review and are being reframed accordingly. Research supervision as part of such change has rightly come under scrutiny. In Universities in both Australia and the UK, there has been a rapid emergence of professional development programs for research training (Pearson & Cryer, 2001). Australian universities have implemented training and development programs for research supervisors with a view to enhancing student/supervisor relationships, providing forums for discussion and closer reflection on supervisory practices, as well as the dissemination of the growing body of research on research supervision. Furthermore, systems of supervisor registration and accreditation have been implemented in order to ensure that the most qualified, experienced and research active academics are positioned well for supervision. The role of supervisor is no longer considered a right of the academic but rather a privilege to be earned.

At the university in which I work, my role is to manage and implement training and development programs for supervisors as well as registration and accreditation. Thus, my work includes assisting supervisors to work within new research training frameworks. Change is very much a part of the larger spaces in which we roam but scrutiny is needed of the seemingly diminishing spaces for supervision. While policy development surrounding research supervision must make both students and supervisors more accountable and progress more transparent, more time and space for supervision needs to be cleared within academic work plans and agreements. As Martin (1999) clearly argues changes in academic work are necessary, even praiseworthy, but often come at a price. We need to consider the changes in the research training context in terms of the cost related to supervision, learning and curriculum, and resist reductionist approaches to research supervision.

PERFORMANCE DRIVEN SPACES: THE RESEARCH STAGE

The university context in relation to research education Australia is now one in which performance is paramount. What constitutes performance has been defined more tightly than ever before as a result of the national emphasis on the knowledge economy and innovation as manifested in the White Paper *Knowledge and Innovation: A Policy Statement on Research and Research Training* (1999) and the Howard Government's *Backing Australia's Ability* initiative. The White paper gave birth to two performances based schemes for research education: the Institutional Grants Scheme (IGS) and the Research Training Scheme (RTS). Both are administered by a Federal body namely the Department of Education, Science and Training (DEST). The schemes are geared to strengthening the research and research training system in Australia. This chapter focuses on the impact of the RTS in terms of research supervision and the research experience.

Australian universities now reside in, and must abide by, a pressurised context of performativity. We are not alone in this respect. We, alongside our students, are told that we have become knowledge workers within a research performance economy. No longer is knowledge simply viewed in epistemological terms but it is now rather framed more and more in terms of outputs. Knowledge, in this context, is both a key input and central output, and therefore managed within institutional systems and processes (Green & Usher, 2003). Not surprisingly we are moving towards what could be termed the "rational" university, that is, the university based on principles of pragmatic managerialism where outputs are commodities for which the most funding can be gained. Does that then mean we are moving toward "rational" supervision? If so, just what does that entail? What is the space for supervision in contested spaces where performativity reigns?

Universities in Australia, under the regime of the Research Training Scheme (RTS), which began in September 2000, are rewarded fiscally through a national competitive scheme where allocations are made through a formula comprising completions (50%), research income (40%) and publications (10%) (Research Training Scheme: Guidelines for 2002, p. 6). Performance gains are tallied, and it is upon this basis that further research degree places (other than full fee) are allocated to universities. The given number of available research places is now restricted such that universities operate largely on a competitive basis with respect to postgraduate students and where success is dependent on timely completions.

Spaces for research degrees within universities are diminishing across the board. With places allocated to universities according to performance and with universities internally allocating places using the same formula, places are limited. With rationed places we can no longer enrol all applicants who meet entry requirements and who we can supervise. Research places for candidates, other than full fee, are far fewer and located more strictly within acknowledged research concentrations or areas of recognised research excellence. Once a student is enrolled under the RTS the pressure to perform begins in earnest. If students fail to meet progress standards and deviate from closely, monitored and managed schedules, valuable research places may be lost in the future. Should students miss final deadlines and fail to submit the thesis (or equivalent) on time, then they fall out of the RTS and incur full fees. Those who cannot afford fees may well drop out completely.

The impact of the Research Training Scheme on Australian universities cannot be understated. In terms of supervision the message is stark and clear. In this context of performativity where performance is tied to funding (Lyotard, 1984), those of us involved in supervision are under pressure to create performances in which timely completions are on central stage. Timeliness is now formulaic. The RTS decrees that a research degree must be completed in a given time frame (4 years full time, or equivalent, for doctorates; 2 years full time, or equivalent, for Masters) (Research Training Scheme: Guidelines for 2002, p. 4). Such time frames are actually far more restrictive than they first appear. The space for supervisors to work with students to produce a thesis (or equivalent) is much changed. We perhaps now find ourselves positioned as performance directors in which production times are severely reduced and resources for staging have not increased (or may even be diminished). Furthermore, we are more at the mercy of the audience (the examiners) with respect to the timeliness of their applause and critique. We must now select examiners not only for appropriateness in terms of expertise, standing and fairness, but also their ability to form judgement within a tight and inflexible time frame.

The framework of available research degrees in Australian universities has emerged so that the traditional PhD now takes its place on the list of research degree options along with the likes of professional doctorates, PhD by project, and PhD by publication. For instance, professional doctorates, relatively new in Australia, emerged in the late 1980s and early 1990s in response to internal pressures within academe as well as drivers related to governmental policy and market forces. Given that many doctoral students do not gain (nor aspire to) positions within universities, the notion of doctoral training as preparation for academe has been deemed, for the most part, limited. Consequently, expanded options have emerged, and supervisory practices must thereby shift to meet the various demands of such options. Furthermore, supervisory models are expanding beyond the traditional face-to-face. We are moving from the one-to-one version once commonplace in the social sciences, and the traditional science model of a laboratory based team, to a range of models. These include cluster groups or team supervision (focused on degree type; methodologically centred; and/or discipline based) where some face-to-face occurs but online communication is prevalent, to the more distant models where online communication is paramount due to issues of distance and/or compressed time.

Consequently, the range of dances that we must be able to direct, all at optimum pace, is perhaps a choreographer's dream (many dances, rapid turn-over of work) but potentially a director's nightmare. Directors or rather supervisors need to be not only pragmatic and paced in their supervisory tasks, but familiar with increasingly complex research processes and products, in tune with the concerns of the audience via which these will be assessed, as well as confident and competent in their own substantive areas and methodological inclinations and knowing. Furthermore, supervisors must take into account the push to develop graduate outcomes in terms of broad competencies. In Australia there is a strong emphasis on the need to develop graduate capabilities as part of the total provision of a rich research experience. At my university, for instance, we are committed to the development of graduate capabilities including problem solving, written communication, critical analysis, project management, independent learning, electronic communication skills and so on. If we are truly attempting to foster the development of a broad array of competencies in our research students, it is questionable as to whether a compressed time frame is the way to do so.

The context of performativity in which we find ourselves is not only pushing us to dance a variety of dances at a greater tempo but also leaving us to deal with extended scrutiny. There is always someone looking over our shoulders to judge our progress in terms of backstage preparations, performance and the documentation of such. Initial auditions for entry into the research dance are no longer sufficient for either directors or dancers. Quality assurance measures are now in full force in Australian universities. These occur at both university and faculty level, and involve closer scrutiny of supervisory practices as well as student progress. Arguably greater visibility is much needed and it is time to put supervision under the spotlight. The focus on progress has pushed many of us into project management with our candidates where projected timelines are documented, followed as closely as possible and reviewed constantly. Our supervisory practices involve close monitoring of progress through progress reports, seminar and conference presentations, thesis groups, and review panels. At policy level "at risk" procedures have been revised so that a harder line is taken when candidates fail to show adequate progress. Whether or not this elicits increased understanding of the needs of supervisors and candidates remains a point for conjecture, but the focus on such reporting is clearly on timely completion and involves greater accountability for all involved. While such developments can be seen to constitute positive steps towards greater student support and more focused candidatures, the impact on the relational not to mention the intellectual aspects of supervision remain to be seen. Thus the dances that we direct are conditional upon the approval of the producer well beyond the audition phase. Where performance is deemed as being too slow and at risk of missing the designated first night (date of submission), the tempo must be increased and the director must push dancers harder in their paces if a stage presence is to continue.

All of this confronts us within diminishing spaces or on a diminishing stage. While on one hand, we might feel the need for stretching exercises so that we can be flexible enough to perform this expanding range of tasks in a growing repertoire of supervisory models, we cannot stretch at all well when the time and space is so compressed. It seems that we are being pressured more and more to take a more "rational" (or rationed) approach to supervision at the very time that this already complex task is becoming even more complex.

RELATIONAL, INTELLECTUAL AND PHYSICAL SPACE FOR SUPERVSION

Let us now consider the space afforded to research supervision, and the impact for supervisory practices and learning. Such space can be seen to comprise three main elements: relational, intellectual and physical spaces (RIP). It is argued that these spaces are being compromised in the push for timely performances on a tightly bounded stage where performers are prepared to dance in a lockstep manner without room for improvisation or deviation from the choreographer's score. For ease of discussion each space is discussed in turn, although clearly they are inextricably entwined.

Relational Space

Relational space for the purposes of this chapter is taken to refer to the time and place in which the complexities surrounding the student/supervisor relationship emerge, shift and even play havoc during the research journey. It has long been established in the literature on postgraduate education that the supervisory relationship is the most important aspect of postgraduate research (Laske & Zuber-Skerritt, 1996). It follows, therefore, that the relational space for supervision is crucial.

The notion of relational space for supervision is based on the assumption that supervision is social practice. In order to consider this assumption further, we need to first examine the term practice. Years ago Scribner and Cole (1981) defined practice as "a recurrent, goal-directed sequence of activities using a particular technology and particular systems of knowledge." While they applied this definition to notions of literacy, it is perhaps useful to do so similarly with supervision. If we think of supervision, the goal directed aspect of the definition works well, as does the elements of technology and knowledge especially when we acknowledge that such elements are dynamic and multiple. These elements will be discussed within the sections on intellectual and physical space. However, this definition, like the RTS, needs then to be located in terms of the social to take into account the relational elements of supervision. The complexities inherent in the delicate relationship between supervisor and student cannot be ignored. We need to situate supervisory practices as part of wider social practices, that position people in various roles and networks within which these roles operate, that rely on given discourse practices, and that occur in relation to cultural practices in specific settings spanning the university, the workplace and industry. Learning how to be, to produce and to grow as researchers within such supervisory relations takes time and space as ways of working develop, expectations are clarified, progress occurs and feedback is given.

By focussing on the relational elements of supervision, we are forced to acknowledge, at the very least, the complexities that exist within the relationship between student and supervisor. The relationship brings together an array of individual characteristics, theoretical positions and approaches to research, worldviews and interests, as well as varying commitments within work and beyond, and ways of managing such pressures in the pursuit of a higher degree.

As Delamont et al. (1998) note, supervision involves the creation of a delicate balance between dominating and neglecting the student's research. Using contrastive rhetoric, the researchers explore the accounts of supervisors about their own experiences as doctoral students with their current practices as supervisors themselves. Reflections by supervisors on their current practices reveal a tension between the need to guide and structure doctoral work and the desire to allow students to be autonomous researchers. This balance would seem to be now under greater threat as research degrees become more highly and more explicitly structured into bounded time frames.

The student-supervisor relationship is often an ambiguous one in which "zones of uncertainty" (Laske & Zuber-Skerritt, 1996) exist. Such zones may be found with respect to organisational structures, economic support systems, political orientation, research culture, the research question, the student, and the supervisor. These zones cannot be clearly defined "because they are experienced and conceived differently by different people. What might be clear and certain for a supervisor may be totally unknown and uncertain for a student" (Laske & Zuber-Skerritt, 1996). Arguably such zones are now expanded further as we work together across research degree frameworks across disciplines across the globe. Laske and Zuber-Skerritt conclude that students need to experience the processes involved in doing a research degree before they know which questions to ask. In a sense students need to "live" the expectations of the supervisor in order to know what issues to discuss and what support to request. Yet the relationship is not one sided. Students come with their own expectations of their supervisors. "Students expect their supervisors to be strongly relationship oriented, whilst supervisors typically show task-related and more impersonal expectations" (Laske & Zuber-Skerritt, 1996). These expectations are potentially problematic in that students want support and assistance but also seek autonomy and independence. Similarly supervisors want students to adapt to "the rules of the game" and proceed accordingly without too many demands, but expect intelligent, creative and innovative work. Clearly not all supervisory relations work well. Conflicts arise often out of differences in expectations. Time is needed to work through to clarity of expectations and to find a working relationship that suits the various needs, wants, personalities and quirks of those involved. Sometimes a change in supervisory arrangements is necessary. Regardless, the space needed for such essential growth of this delicate relationship cannot be underestimated.

Power differentials exist within the supervisory relationship. Clearly the student-supervisor relationship is one where the power balance leans towards to the supervisor. Students, no matter how accustomed to speaking up, may still feel reluctant to make too many demands on their supervisors. As time goes on, and the relationship strengthens and student knowledge of the substantive area and the research approach grows, students are often better placed to speak up and state their views, challenge the status quo and suggest their own ways of framing their work. Such growth takes time and needs to be nurtured. This is not simple at the best of times but is it even more problematic and involved when time frames are rushed and the performance is choreographed tightly.

Supervision is not strictly academic (Grant, 2001) but rather revolves around a complex relationship between student and supervisor. Grant urges us to see supervision in its complexity and to consider issues often left unspoken. She notes that supervisory codes of conduct tend to silence the complexities within supervision by failing to acknowledge the issues of power, desire and difference. Her paper highlights that transgressions, often rendered unspeakable, do occur at times within supervisory practices. While these are sometimes intolerable, liaisons do occur. Although full discussion of such issues is not possible here, by raising rather than silencing issues related to power, desire and difference, Grant acknowledges the emotional elements involved in the supervisory relationship. Her work connects strongly with the earlier work of Lee and Williams (1999) who forcefully argue that the emotional, and the irrational should be acknowledged, studied and theorised further in order for us to learn more about the complexities surrounding doctoral work.

In their highly regarded work, entitled *Forged in Fire,* Lee and Williams (1999), examine the role of emotions and the irrational in doctoral supervision. In particular they uncover the phenomenon of trauma and distress, and argue strongly that emotion should not be silenced nor rationalised away. Instead they take the view that the emotional and irrational dimensions are a necessary part and effect of the production of a "licensed independent scholar." Lee and Williams (1999) posit, "As student-subjects are produced through complex relations of power-desire-knowledge, in which the supervisory relation is implicated, the process of attaining an academic-disciplinary identity/position is often characterised by trauma, contradiction and ambivalence." The researchers collected stories of being supervised from a group of elite scholars. Stories of the desire to become "the other" (the scholar) were marked by feelings of pleasure but also trauma, and great emotional turmoil. Such complexities will persist even if policy makers attempt to silence rather than recognise these emotive dimensions.

The importance of relational space for supervision cannot be overemphasised. Supervision is a social practice, and therefore takes up time in a socially located place or space. Relational space requires physical space, particularly the face-to-face. Arguably such space, which takes time to develop, is vital to the growth of intellectual space on the current research stage. The relationship between dancer (student) and director (supervisor) is crucial if the dance is to take place in ways (significant, original, intellectual contribution) deemed appropriate by the audience (examiners) and producers (funding bodies, universities). Thus, while it is evident that the three spaces are not necessarily equal in terms of size or significance, they are indeed interdependent and crucial to overall performance.

The relational space needed for supervision is both literal and metaphorical. If we fail to acknowledge the complexity of supervision and reduce the space given to relational elements, we run the risk of losing sight of arguably the most crucial and challenging aspect of supervision. Relational space sets the scene for, and provides momentum to, the intellectual aspects of a higher degree. We now move to consideration of the space for intellectualism within supervision.

Intellectual Space

Although supervision is not strictly academic as noted above, it does centre on intellectualism. Opportunity for intellectual space is another crucial aspect of supervisory practice. While the research degree may seem to be choreographed to a point, the role of improvisation cannot be forgotten. The audience seeks a level of improvisation on stage in order for the performance to be deemed as new or significant. In creating performances anew, dancers, supported by their directors, must dance in ways never before witnessed by the external audience.

While it is argued that relational space is a key aspect of supervision and seems to be very much under threat within the RTS, the remaining space for the intellectual cannot be ignored. The pursuit of a research degree is an intellectual activity, and a key role of the supervisor is to stimulate, grow and review the intellectual within the candidate per se but also as evidenced by the visual durable product that is the examined outcome. It follows that supervisory practices must provide space for the intellectual.

Intellectual space is a shorthand term for the time and place needed for the generation of fresh ideas. Such space is not merely seen here in terms of the familiarisation with or consideration or critique of the literature pertaining to a given field (or across fields) but rather is taken to include notions, ideas or knowing that occur anew. Given that research degrees are judged on such criteria as the "originality" and/or "significance" of the contribution in terms of knowledge, the spaces for intellectualism within supervision is crucial.

Bowden and Marton (1998) emphasise that the object of learning within research is more constrained (in that we aim for something new) but that the acts of learning are less so (in terms of time). The authors perceive the fruits of research as taking time—years even decades. "Trying to find out something that nobody has found previously is different from trying to find out what somebody else has found out earlier" (Bowden & Martin, 1998). It follows, therefore, that the learning that is afforded by the pursuit of a research degree needs time. The curriculum design that supports such learning must provide space for the generation of fresh insights; the testing of new discoveries against the backdrop of established knowledge; as well as reflection in terms of ideas, research practices and connectivity between the theoretical and the practical. Fresh insights cannot be tightly choreographed. Intellectual growth that can be judged as being a "substantial" or "significant" contribution requires time for the exploration of theoretical frames, for the pursuit of research activity, to seek audiences for critical feedback and dialogue, and to return to seminal ideas, but also to move outside the box of traditionally located or acknowledged ideas in order to find knowing anew.

Supervision, therefore, must be viewed as a pedagogical pursuit in so far as it is a complex task involving the production of meaning, bringing together, as Smith (2001) reminds us, the intersections between teaching, learning and knowledge production. Supervision is not about the transmission of knowledge but rather the cultivation of a context in which new ways of knowing or seeing are found.

As noted earlier in this chapter, our research degree frameworks are expanding and the roles of supervisors are changing and becoming increasingly complex. Once supervision was based primarily upon substantive knowledge. The supervisor was expected to be expert in terms of content. However, more and more supervision demands an understanding of the candidate's discipline(s), methodology, and/or the practical field. A supervisory team that spans such areas is needed. Arguably, the team would consist of at least an academic located within the discipline(s) and/or methodology, as well as a practitioner in the field. Thus, space is needed for supervision across academe and the workplace.

The space for supervision must allow for reflexivity whereby candidates are given opportunity to critically reflect on their ongoing contribution to knowledge. As *The Reflective Spin* (1999), edited by Chen and Van Maanen, reveals, space for reflective practice, on the part of both teachers (or supervisors) and learners (or candidates) is a vital part of the learning journey within higher education. Time and place(s) for reflexivity is essential for the growth of ideas, the move towards theory, the emergence of a research design and appropriate approach, and the transformation of action. Diminished space for reflective practice will affect intellectual growth. We must find more rather than less space for such practices.

The threat of diminished space for the pursuit of a higher degree has potential to constrict the range of research approaches for our students. This has grave implications for the ways in which knowledge is produced. For instance, in a full time Masters program of 18 months, there is little opportunity for the pursuit of action research in that there is no time for numerous cycles of inquiry and reflection unless the candidate studies on a part time basis. Longitudinal studies based on ethnographic approaches may no longer be deemed suitable methodologies, especially for Masters students and full time doctoral work where timelines are very much constrained. The room for methodological choice is a vital aspect of connecting research questions with appropriate research approaches and is in danger of compromise if the space for growth of the intellectual is diminished.

Physical Space

The physical space required for supervision can be seen as being quite simple. Essentially what is needed is opportunity for face-to-face meetings between supervisor(s) and candidate. While contact at a distance may sustain for a period, face-to-face meetings between supervisor and student are crucial in terms of motivation, clarity of expectations and feedback, and planning for future work. The face-to-face is vital in terms of relational space and the need to reduce where possible zones of uncertainty. McWilliam and Taylor (2000) note that it is face-to-face contact that creates the momentum needed to motivate and drive students on to the next stage. The face-to-face requires time and place, which implies fiscal resources that enable these two elements to coincide. Other physical resources refer to computers, books, equipment, communication channels such as email access, a place for students to work (only for full time students), financial support via scholarships, conference attendance and so on- all of which are viewed as adding to the research experience. The degree to which these resources are available varies from university to university, and from candidate to candidate. Just who takes responsibility for the provision of such resources is an issue best left to discussions outside of those centring on supervision. This is not to de-emphasize the input of these resources but rather to focus on physical space for supervision comes down to the face-to-face. Where the space for the face-to-face is limited, other resources in particular electronic communication such as email, become crucial. Now that candidates are under pressure to stage timely performances the physical space for supervision must be preserved. Our academic work plans must provide room for appropriate space for supervision, just as our schemes for promotion need to raise the profile of such work. The relationality and the intellectualism involved in supervision are just as reliant upon the physical space for supervision as they are upon each other.

DIMINISHING SPACES: A FINAL NOTE

This chapter has focused on the impact of the changing research training scene in Australia on supervision from the vantage point of supervisors, candidates and those involved in supervisor development and training. The notion of regulated space for the completion of a research degree creates a context where the driver is the RTS and the formulaic is seen to equate with quality and accountability. The question of what constitutes the new spatial dimension for research supervision, or the "new space" for supervision, has been examined. The argument has been made that spaces for supervision on the research stage are diminishing. The relational, intellectual and physical spaces (RIP) in which research supervision resides (or is bound) are being threatened in ways that silence the complexities surrounding research supervision. The interconnectedness of these spaces has been argued, and while they may differ in terms of size and/or significance, each is a vital part of the overall space for supervision. Arguably these different spaces are essential for the generation of different but complementary forms of knowledge that contribute significantly to the overall performance on the research stage. If the director's space to work with the dancer is constrained, the development of the dance repertoire, and the inherent skills, as well the improvisation involved, is put at risk. Should the physical space allowed for preparation is limited, lack of clarification is a potential problem that might ultimately constrain performance.

The compression of the time and place available for supervisory practices pushes supervision into strictly paced, tight spaces, and fails to take into account that supervision is a social as well as a pedagogical practice. The new context in which we supervise silences the multiplicity of practices within our midst that fall under the heading of research supervision and neglects to acknowledge that supervisory practices vary across personalities, traditions, management styles, distance and access, disciplines, research degree frameworks, and so on.

The imperative is for us to further theorise the pedagogy of supervision. The cry to see supervision as pedagogy and to theorise supervision into a more sophisticated framing is not new (Lee & Green, 1995). Now that performativity guides what counts, we must face the challenge and the need to complicate, rather than simplify how we view and make space for research supervision. The stage may well be set for the next performances on our dance programs but the backstage discussions must continue in earnest if we are to protect the spaces needed for the supervisory practices supporting research degree experiences and outcomes that count from the view of the external audience but also from both supervisor and student viewpoints.

REFERENCES

Bowden, J., & Marton, F. (1998). *The university of learning: Beyond quality and competence in higher education.* London: Kogan Page.

Chen, A., & Van Maanen, J. (1999). (Ed.s). *The reflective spin: Case studies of teachers in higher education transforming action.* Singapore: World Scientific.

Delamont, S., Parry, O. & Atkinson, P. (1998). Creating a delicate balance: The doctoral supervisor's dilemmas. *Teaching in Higher Education, 3*(2), 157–172.

Department of Education, Science and Training, (2002). *Research training scheme: Guidelines for 2002.* Canberra: Department of Education, Science and Training.

Grant, B. (2001). Dirty work: 'A code for supervision' read against the grain. In A. Bartlett & G. Mercer (Eds.), *Postgraduate research supervision: Transforming (R)elations.* New York: Peter Lang.

Green, P., & Usher, R. (2003). Fast supervision: Changing supervisory practice in changing times." *Studies in Continuing Education.*

Kelly, R., & Ling, P. (2001). A posttraditional supervisor-supervisee relationship. In A. Bartlett & G. Mercer (Eds), *Postgraduate research supervision: Transforming (R)elations.* Peter Lang Publishing Inc: New York.

Kemp, D. (1999). *Knowledge and innovation: A policy statement on research and research training.* Canberra: Commonwealth of Australia.

Laske, S., & Zuber-Skerritt, O. (1996). Frameworks for postgraduate research and supervision: An overview. In O. Zuber-Skerritt (Ed.), *Frameworks for postgraduate education.* Lismore, NSW: Southern Cross University Press.

Lee, A., & Green, B. (1995). Introduction: Postgraduate studies/post graduate pedagogy? *The Australian Universities' Review, 38*(2), 2–4.

Lee, A., & Williams, C. (1999). 'Forged in fire': Narratives of trauma in PhD supervision. *Southern Review, 31*(1), 6–26.

Lyotard, J. F. (1984). *The postmodern condition: A report on knowledge.* Translated from the French by Massumi, G., & Bennington, B. Minneapolis, MN: University of Minnesota Press.

Martin, E. (1999). *Changing academic work: Developing the learning university.* Buckingham: The Society for Research into Higher Education and Open University Press.

McWilliam, E., & Taylor, P. (2000). *Reconciling rapidity and rigour: Strategies and structures for the next generation of professional doctorates.* Paper presented at the Third International Professional Doctorates Conference, Armidale, Australia, September.

Pearson, M., & Cryer, P. (2001). 'Training the trainers': Current initiatives in enhancing research supervision in australia and the UK." *Herdsa News,* April.

Scribner, S., & Cole, M. (1981). *The psychology of literacy.* Cambridge: Harvard University Press.

Smith, B. (2001). (Re)Framing research degree supervision as pedagogy. In A. Bartlett & G. Mercer (Eds.), *Postgraduate research supervision: Transforming (R)elations.* Peter Lang Publishing Inc: New York.

Usher, R. (2002). Subject and object in the contemporary university: An exercise in autobiography. In P. Green (Ed.), *Slices of life: Qualitative research snapshots.* Melbourne: RMIT Press.

CHAPTER 12

VIRTUAL RESEARCH PRACTICES AND PHENOMENOLOGIES OF THE INTERNET

Robyn Barnacle

While a great deal has been written about on-line learning in education, very little attention has been devoted to its role in postgraduate research degrees—PhDs, Professional Doctorates and Masters by research[1]. For research students, however, whose program of study is largely self-defined and for whom keeping pace with the latest ideas both nationally and internationally is pivotal, new information and communication technologies (ICTs), principally the Internet, have much to offer. The use of such technologies has, and will, change the nature of the research degree, as use of the Internet, in particular, re-orders our relation to each other and to knowledge. This re-ordering has significant spatial dimensions.

Most obviously, being located on-campus, co-located with other researchers, is no longer necessary. This is demonstrated by the increasing numbers of research degree students enrolled part-time, both in Australia and abroad (Barnacle, 2002; Evans, 2002). Such students are commonly located off-campus and regular contact with supervisors and access to libraries and other information resources can be achieved on-line. Even for traditional full-time on-campus research students, however, ICTs have

Space, Curriculum, and Learning, pages 173–188

the potential to facilitate access to an international research community way beyond the scope of what was traditionally available within the confines of a single University campus.

The Internet has also had other, perhaps less tangible, spatial impacts. Most critically, the Internet alters the learners' position in respect to knowledge. While the Internet is not solely the cause of this, nor its only manifestation, it is significant in that it embodies a transformation of traditional models of knowledge production and dissemination, in the process undermining traditional, hierarchical, models of disciplinary and institutional authority. Universities and disciplines are no longer the sole gatekeepers of what counts as knowledge.

This chapter examines the spatial re-ordering that is engendered by use of the Internet and its impact on the research degree. This impact is particularly evident in relation to the social dimension of learning—often seen as critical to the research student experience. What digital technologies make possible in terms of email and library resources, some may feel they also take away in the form of the human interaction that has been traditionally associated with a vibrant research culture. The fracturing of the traditional researcher—research community nexus that is facilitated by ICTs challenges the way in which interpersonal relationships and communities and, therefore, knowledge transfer, are conceived. That is, it challenges the centrality of spatio-temporal proximity within human relations and its significance to the induction of students into fields of knowledge.

Exploring issues associated with technology and learning can be enriched through the insights of phenomenology. The traditional focus within phenomenology on praxis and embodiment make it particularly relevant for addressing issues related to the impact of technology on our lives. By comparing the approach of two phenomenologists who deal with this subject, Dreyfus (1996) and Ihde (1993), I explore what a phenomenological perspective has to offer for understanding the impact of ICTs on research practice.

THE RESEARCH STUDENT EXPERIENCE

Globalization and the emergence of new ICTs, particularly the Internet, are having considerable impact on education and traditional models of learning. As Burbules (2000a) suggests, "...educational institutions, at all levels from primary to higher education, are trying to establish themselves as communities in and of the Internet." By impacting on conventional models of both time and space, globalization and the Internet are transforming traditional conceptions of learning (Edwards & Usher, 2000; Peters, 1996). As Edwards and Usher (1998) point out, "the very notion of a 'course' that takes place at a fixed time with pre-defined starting and end points and within fixed designated spaces is becoming increasingly problematic as space is restructured and time transformed with the impact of globalization."

While the meaning of globalization itself remains contentious, it can be characterized by two key factors: a mobilization of internationalizing processes, which are said to create and sustain globalizing networks, flows and impacts; and the means by which this is largely facilitated and sustained. As Manovich (2001) proclaims, "...today we are in the middle of a new media revolution—the shift of all culture to computer-mediated forms of production, distribution, and communication. The link between new media forms like the Internet and globalization is intimate and inextricable." Hence, for Burbules (2000b): "the Internet is thus both one of the chief manifestations of globalization, and one of its primary causes." Together, globalization and ICTs have considerable cultural, political and economic impacts. Spatial and temporal barriers are transformed, enabling a global flow of information, not from one place to another, which implies a privileging of origin and destination, but instead, through the creation of virtual spaces that can, in principle at least, be accessed from anywhere, and as a consequence, exist everywhere. Notions of locality and the local become transformed, and with them, traditional conceptions of education as what Edwards and Usher (1998) describe as "fixed curricular time and space."

As Urry (1998) and others (Brown, 2001) suggest, globalization can be understood as simultaneously both undermining and privileging the status of the local in relation to the global. What were once considered constraints or advantages of geographical location have become transformed. So much so, that the traditional model of the research degree can be seen as clearly derived from a pre-digital age. For, as Pearson (1999) suggests, "the conventional picture of doctoral education which assumes an on-campus, full-time student experience, with socialization arising formally and informally through interaction with the supervisor(s) and other academics in a university department ...is increasingly recognized as problematic." ICTs promote heterogeneity in the places within which it is possible to do research and from which it is possible to connect with a research community. In contrast to the traditional model of the research degree, as Pearson (1999) suggests, communication between academics and scholars is "...being facilitated by the increased availability and sophistication of electronic communications." What then is the impact of how and where research is conducted on the research degree experience and what do virtual research communities have to offer in relation to the traditional face-to-face model? More broadly, how are ICTs altering research practices and are they enhancing or detracting from the research student experience?

RESEARCH ON THE INTERNET

While interest in the role of ICTs in education has been developing in recent years, less attention has been given to their role within research degrees. Most notably, literature on on-line and distance learning tends to focus on the role of ICTs in the delivery of under-graduate or post-graduate course work degrees. Where there is consideration of web-based delivery in the context of research degrees the focus is largely on supplementary 'core-training' programs such as generic skills and research methods. Barrett and Lally (2000), for example, report on an interactive multimedia (IMM) environment designed to support research methods training for distance students (using a CD-ROM package called Signposts for Educational Research). In their evaluation of the program they raise a number of issues that are salient for questions regarding the role of ICTs within research degrees generally. They question, for example, "...whether research students can 'become critical' in a multimedia environment." Barrett and Lally (2000) were concerned that students' use of the program was highly instrumental as it was treated simply as a data resource, useful for downloading bibliographical and other material. While the authors concluded that the IMM program that they were evaluating was able to support both instrumental and reflective approaches to learning, the issue regarding the kind of learning that is promoted by electronic learning environments remains.

Assessing such issues in relation to the unstructured or non-curriculum driven use of ICTs in learning is more difficult. Use of the Internet by research students can be understood as unstructured in as, for the most part, such degrees are characterized by little or no set curriculum. Instead, under the guidance of a supervisor or supervisors, research students define the content, methods and outcomes of their research themselves. As a consequence, such students actively construct a program of study through the research process itself. In this sense, the course of study of a research student can be understood as emerging through the very act of doing research. Compared to a course-work degree, in which the program of study is established in advance, research degrees are relatively unstructured and emergent. My interest is in the role of ICTs as enablers of this emergent feature of the research degree. How might the use of ICTs facilitate, alter or impact on the kind of research practices, and ultimately the learning experience, of research students?

In thinking through this question there are two intersecting yet distinct themes that need to be considered. The first centres on how ICTs might aid in the creation and execution of a research program, or the ways in which the research itself is carried out. The second is the role of ICTs in facilitating communication between the student and supervisor(s) couplet, and beyond that, to an extended community of other scholars and researchers. Both of these elements impact on the research student experience: on the kind of activities that characterize the research experience and the practices for doing research that are deployed.

Attending to the question of how the Internet impacts on learning reveals what makes the Internet different than other developments in learning like off-campus and distance models. The Internet is similar to the distance-learning model as it breaks down the need for physical co-location between student and supervisor. But the spatial re-ordering of the Internet is not just limited to the physical. More significantly, in relation to traditional models of information dissemination and exchange, like the library, lecture hall, or course reader, the Internet re-casts the learners' relationship to knowledge. The hyperlink promotes the individual, errant learner, in a continuous flow of discovery, destabilizing traditional, hierarchical models of disciplinary and institutional authority. On the Internet, the link has become the new currency for ordering knowledge, with search engines, in particular, determining value on the basis of "hits" (Brown, 2001). This challenges the economy of the expert, which traditionally ordered what did and did not count as knowledge. Added to this, real-time interactivity mechanisms such as email and chat rooms enable opportunities for exchange between physically dispersed yet temporally simultaneous individuals in spontaneous, or unauthorized, learning forums. If the classroom is the web then there can be little determination—in principle[2]—of who can participate and what can be discussed. The spatial relations transformed by the Internet impact upon knowledge and identity, rather than being just physical—as hierarchies of knowledge are conceived vertically, the hyperlink situates knowledge horizontally; it becomes a depthless flow of information.

Clearly, the spatial re-ordering that is promoted by the Internet has implications for all types of learning and education delivery. The Internet might be thought of to have particular relevance for the research degree, however, as its features seem particularly conducive to the individualized and emergent nature of such degrees, having the potential to compound and enrich what already makes them unique and, presumably, appealing. For the research student navigating the hyperlink might be understood as a natural extension of traditional research methods like library searches. The challenge would be to become abreast of a knowledge field that, due to the proliferation of material that appears on the Internet, and through publication in general, is no longer containable. This in itself has the potential to transform research degree practices, as the geographical and other factors that traditionally might have limited the scope of a study may no longer apply. Unfortunately, however, little data exists to show how research students are actually utilizing ICTs. Preliminary indications at my university demonstrate that, if nothing else, electronic communications devices like email and on-line library resources, specifically serial collections, are being heavily utilized by research students. But it is unlikely that this is the full picture, particularly considering the range of features that ICTs make available. Further research is required to ascertain the extent of ICT use amongst research students and the features that are being utilized.

VIRTUAL RESEARCH COMMUNITIES

One area where the Internet might be thought of as most fruitful as an enabler of the emergent nature of research degrees is in its potential to facilitate virtual research communities, or on-line learning forums. In principle, ICTs allow students to be spatially distributed—yet simultaneously electronically connected—to other researchers throughout the world and their ideas. This is the promise of the 'global community', the many-to-many, that ICTs make possible. Examples of electronically mediated communities include email discussion lists, of which there are thousands on a myriad of topics (for example, see www.fibreculture.org), and Internet chat rooms, both of which allow for "real-time" discussion between multiple participants in different geographical locations. Other features like personal web sites, or blogs—web-based diaries or logs (see www.blogger.com) provide almost instantaneous personal web pages that can be uploaded multiple times in the same day, providing both intimacy and immediacy of communication.

In terms of much of the literature on the research student experience, however, the link between conceptions of a good research environment and the traditional research degree model remain strong. Such conceptions emphasise notions of co-location and critical mass, where a vibrant research community is almost always equated with face-to-face relationships (Lovitts & Nelson, 2000; de Valero, 2001). Lovitts and Nelson (2000), for example, address factors that contribute to non-completion amongst a sample of research students from two universities in the USA. They found, amongst other things: "...a high correlation between integration into a department's social and professional life (becoming part of the community) and successful completion of the Ph.D." A perception of the intrinsic value of critical mass has also become evident in Government policy on higher education both in Australia and abroad (Kemp, 1999a, 1999b; Harris, 1996). The Harris (1996) report claims, for example, that "there is a strong argument that postgraduate research education is likely to be delivered most effectively in the context of a critical mass of research activity."

But the critical mass model also has its critics. Delamont et al. (1997) are critical of what they see as the "extent to which the 'critical mass' assumption has become taken for granted in UK higher education" and challenge such conceptions for failing to address disciplinary differences. They claim that models of critical mass are premised on the lab-based research practices of the natural and theoretical sciences and are consequently much harder to achieve, and perhaps also inappropriate, for the humanities and social sciences.

The notion of critical mass also raises another issue, however, which Delamont et al. do not address: that of part time and off-campus research students. How might critical mass be achievable for these students? A related issue is addressed by Brehony and Deem (2000) in their work that explores how research cultures are accessed and experienced by different types of students within the social sciences i.e.; full-time/part-time, local/international, male/female. They (2000) note that access to research cultures often occurs serendipitously for part time students, and that, overall: "international students and part-time students ...found much more difficulty in becoming part of peer student cultures and academic cultures than did home and full-time students."

While this conclusion is no-doubt valid, like the other studies mentioned above, however, Deem and Brehony do not address the role of ICTs on the research student experience and the role that the Internet and other mechanisms might play in facilitating access to a research community. The use of on-line discussion forums and the like within research practice has the potential to inform a radically revised notion of critical mass, albeit stripped of its face-to-face element. For research candidates enrolled in small institutions, in particular, or where the research community around a particular area is small, virtual communities must surely be attractive. However, even large established research institutions may struggle to maintain their traditional models. Why, for example, would research students be compelled to attend the local seminar series when they can go on-line with cutting edge researchers in the field from around the world? While one must be cautious about suggesting the inevitability of a binary model—virtual versus real—these may become very real issues for those managing research degree programs.

Introducing the notion of a digitally or electronically mediated 'virtual' research community is radical as it goes to the heart of how we conceive of communities, and of ourselves. While the notion of what constitutes community has itself been contentious since well before the advent of the Internet (see Burbules, 2000a), the global dimensions of the Internet raise new issues. While most educational commentators acknowledge the opportunities offered by virtual communities, their enthusiasm is often tempered (Sakamoto, 2002; Corry & Tu, 2001; Chen & Hung, 2002). There seems to be a common sentiment in relation to virtual communities that they are secondary to face-to-face communities when it comes to opportunities for learning.

One of the key reasons for this centres on conceptions of what constitutes the social, and thereby, of what it means to be human. Concerns centre on the abstraction of virtual communities from the social fabric of traditional human relations. Yet as Lewis (in Brown, 2001) points out, such views can be "too reliant on a conception of community which imagines online interactions as somehow abstracted from broader sets of social relations." The notion of the virtual is misleading if it is taken to mean the extreme opposite of the non-virtual or real, since we are physical and social beings that cannot exist solely on-line. The Internet is a human product, after all, and therefore inextricable from the social. But while the Internet is a human product, it is one that far exceeds the scale of the human—it is big, amorphous and impossible to comprehend in its totality. The Internet is grounded in the social but it also transforms the social. How might we understand the sort of human relations and practices that this promotes?

PHENOMENOLOGIES OF TECHNOLOGY

The question of the impact of technology on what it means to be human has been an abiding concern within the philosophy of phenomenology, particularly through the work of Heidegger. There are important differences, however, amongst contemporary phenomenologists as to how the legacy of Heidegger's critique of technology is interpreted in relation to contemporary technologies. Examining some of the differences between these thinkers reveals useful insights into how the role of the Internet might be understood in promoting or supporting learning communities. For an understanding of these differences one must begin with Heidegger.

Significant spatial metaphors are at work in Heidegger's thinking about the nature of human existence, *Dasein* (literally 'there-being') in his early work, and *dwelling* in his later work. Authenticity is linked to notions of nearness and closeness and the lack of authenticity to its opposite, distance and forgetting. For Heidegger, the nexus between man (sic) and world is primordial, as human being can only be truly understood in terms of proximity to Being. It is in this context that Heidegger's critique of technology needs to be understood. This critique aims to preserve or reaffirm the closeness of humans and world, which Heidegger sees as under threat by technology, or more specifically, *enframing*, the reduction of all things to useability or resource. In Heidegger's early work on technology, such as *Being and Time*, he was concerned that its effect was to dis-locate, or alienate and distance *Dasein* from its more essential relation to things as being-in-the-world. Technological thinking covers up or conceals that relation under the reductionist instrumentalism of thing-as-resource. By the time of *The Question Concerning Technology* (1993 (1954)), however, Heidegger revised his critique of technology such that even the role of a desiring subject, ordering resources for particular ends, albeit instrumental, becomes undermined. Under this later account of technology, order occurs for its own sake, incessantly perpetuated by itself and the logic of increased flexibility and efficiency. The subject as agent becomes subsumed within the all-consuming *gestalt* (or world-revealing) of modern technology.

Dreyfus (1996, 2001) describes this mode of technology as postmodern, and points to the Internet as its ultimate example. In the former, Dreyfus contrasts the cultures of the modern library to that of the new information retrieval practices characterized by the Internet to illustrate his concerns. Whereas the former involves careful selection, classification and permanent collections, the latter is characterized by access to everything, diversification and dynamic collections. Dreyfus (1996) summarizes the post-modern culture of information retrieval: "...the user seeking the information is not a subject who desires a more complete model of the world, but a protean being ready to be opened up to ever new horizons. In short the post-modern human being is not interested in collecting but is constituted through connecting." Dreyfus is concerned by the kind of being that the Internet encourages us to become, or more specifically, that it encourages us to become anything at all. Using the Internet promotes the formation of fractured identities, or selves who are fluid, dynamic and eclectic. "Such people would thrive on having no community and no home sense of self" (Dreyfus, 1996). Dreyfus' archetype is that of the rock group, or what he sees as perpetually ephemeral collections of people, lacking in stability and continuity, in contrast to classical music, and presumably, musicians. While this is a rather clichéd view of rock groups (and carries tones of cultural elitism), Dreyfus' point is that using the Internet promotes a subject who lacks identity and belonging, instead flitting aimlessly from one experience to the next, from hyperlink to hyperlink. This is the seductiveness of the Internet; the promise of unlimited choice, and with that, endless opportunities for self-renewal. But this is dangerous, says Dreyfus (1996), because it "...disaggregates our identities into a contingently built up collection of skills." We forget our role as "world creators" and become, instead, unwitting products—or extensions—of the Internet: nothing more than an aggregation of Internet skills—appendages of the machine, as it were. With that, other ways of being are forgotten. Take, for example, the writing practice that is promoted by the word processor. Dreyfus (1996) warns that "if the tendency to rely completely on the flexibility of technological devices is not resisted, we will be left with only one kind of writing implement promoting one style of practice, namely those of endless transformation and enhancement." The word processor changes the way that we write by, for example, promoting flexibility over decisiveness (as writing can be constantly written over). Email is another example where new writing styles have developed, thereby transforming the formal practices of traditional letter writing. Dreyfus is concerned by the seductiveness of the practices promoted by the Internet, claiming that they undermine our ability to be world revealing, and we therefore lose sight of our potential to create genuinely new ways of being.

Does this mean that Dreyfus is a Luddite, proclaiming that we should abandon all technology? Not exactly, as, following Heidegger, he feels that it is possible to have a positive relation to technology, "but only so long as we maintain skills for disclosing other kinds of worlds" (Dreyfus, 1996). This could take the form of resisting the seductiveness of the word-processor in preference of the pen, for example. There are a number of features of Dreyfus' account of the Internet, however, that are troubling. In particular is his apparent supposition that readers will interpret equating surfing-the-net with being in a rock band pejoratively. Also striking in Dreyfus' account of the Internet is the assumption that the lack of a catalogue or hierarchical system of ordering information makes the Internet necessarily problematic. Dreyfus' concern that the Internet is a leveller of information, creating a lack of differentiation between the significant and trivial, has echoes of a yearning for an authority figure. As Lovink suggests in his review of *On the Internet*, Dreyfus (2002) appears to fear the "tyranny of the digital commons": that the Internet undermines the possibility of a sanctioned, authoritative voice. But Lovink also suggests that Dreyfus' concerns about the Internet are misplaced. For Lovink, Dreyfus' conception of how people use the Internet is clichéd and outmoded, countering that users are now more savvy than they were in the early days of Internet use, better able to navigate their own way on the Internet and able to actively create their own meaning. In line with other commentators on information technology, like Agre (1999), Lovink suggests that it is not the technology that is the problem but that of the values or politics informing its use, such as equity of access.

Ihde's (1993) approach to modern technology provides a useful contrast to that of Dreyfus because, while he shares an affinity with Heidegger's work, he also questions the largely negative status attributed to modern technology. Ihde inquires into why some technologies are treated positively by Heidegger and others negatively, and detects what he calls a strain of 'old fashioned romanticism of a nostalgic sort' in Heidegger's critique. Heidegger's differentiation between 'good' and 'bad' technologies, says Ihde, is informed by a romantic tendency to prefer the smaller simplicity of traditional technologies in contrast to the larger and more complex technologies of the modern era, the preference for the pen over the typewriter being a case in point.

According to Ihde, there are two elements that inform Heidegger's preference for some technologies over others. The first is a preference for "embodiment relations," or "those technologies which express straightforward bodily, perceptual relations with the environment" (Ihde, 1993). In the case of writing, says Ihde, Heidegger feels that human beings 'act through the hand', such that the word and the hand are inextricable (think of the link between speaking and gesture). The problem with typing is that Heidegger sees it as somehow removing the writer from that essential relation. The pen, on the other hand, according to Heidegger's account of equipmentality, "withdraws" in the process of writing and therefore does not obstruct the word and hand nexus. But Ihde sees a limitation in Heidegger's preference for embodiment relations, as a positive relation to technology is only possible with technologies that 'withdraw' in the act of using them. This denies the possibility of technology itself becoming an object with which we can develop a relation, "...and the kind of appreciation and fascination which characterizes much of the experience of modern technologies" (Ihde, 1993). Ihde uses the example of child's toy, such as a spinning top, which itself becomes an object of fascination and delight, and therefore takes on the relation of a 'quasi-other' to the child. Such a relation is not accounted for within Heidegger's account of equipmentality, and it is tempting to speculate to what extent the Internet might also exceed such a model.

The second element that informs Heidegger's determination of "good" and "bad" technologies, according to Ihde, arises from the determination of different technologies as either authentic or unauthentic. Ihde is concerned that this determination is clouded by romanticism. For Heidegger, the pen is more authentic than the typewriter because the world that it "gathers," or brings into being, is conducive to human dwelling, whereas the gathering of the typewriter is that of the enframing worldview of modern technology. Whereas the former is thought of as promoting authenticity the latter is seen as destructive. But for Ihde such divisions conceal a nostalgia and romanticism for the artefacts of previous ages, particularly the architecture of ancient Greece as opposed to that of the moderns. Ihde suggests that Heidegger conveniently de-politicizes older technologies and thereby denies that they too may have had destructive impacts. He does not suggest, however, that we should abandon Heidegger's critique of technology in its entirety.

> What are needed are not a rejection of the deep and essentially phenomenological insights into technology as a culturally embedded phenomenon with its different gestalt features, but a deepening and more complex appreciation of all of the facets of our technologically textured mode of life. And this includes and must include the explicit recognition of both the politics of our

> artefacts, and the demythologization of nostalgic and romantic views of previous times. (Ihde, 1993)

Ihde aims to extract the romanticism from Heidegger's account of technology and question the division it sets up between modern and ancient technology. He calls on us not to abandon Heidegger, but on the contrary, to use the insights of his account of technology to facilitate a better understanding of 'our technologically textured mode of life'. The contrast with Dreyfus could not be greater. Whereas Dreyfus uses Heidegger to argue for an at least partial retreat from such a technological life, Ihde urges the opposite. While Ihde concedes that he is only just in the early stages of working out what this might mean for his own work, what he argues is that modern technology need not be understood as impacting negatively on how we understand ourselves in the world. The danger of technology recognized by Heidegger is not exclusive to our own era, and the positive role Heidegger attributes to ancient technologies in promoting forms of life conducive to human dwelling are also possible through the forms of life promoted by modern technology.

UNDERSTANDING THE ROLE OF THE INTERNET IN PROMOTING RESEARCH COMMUNITIES

It is clear that for Dreyfus the only communities made possible by the Internet are those of a dysfunctional kind, the rock group being the archetype. While it is tempting to attempt to argue the opposite, that the Internet offers the opportunity for truly "functional" communities, one must hesitate about the underlying assumptions required for such a claim. For it raises the question, what is a 'truly functional community'? Are fractious and fleeting identities and relationships necessarily dysfunctional? Or in Dreyfus' model, is the rock group model of human relations really a bad thing? While it is not possible to fully address these questions here, the issues that they raise are pertinent to how electronically mediated communities might be understood. For such issues go to the heart of Heidegger's phenomenological insight: that through technologies worlds emerge, and in these worlds we become in different ways. It is important to be open to what these worlds might mean and not dismiss them in advance based on more established, or perhaps antiquated, models of human relations. Although not informed by a phenomenological approach, this is in many ways what Burbules (2000a) does when he questions the primacy of any single model of what constitutes a good learning community: "if categories like 'learning community' are understood in too homogeneous a manner, then participants with different learning styles or different appetites for affiliation may be left out of them.... The traditional associations of community with proximity, homogeneity, and familiarity can be an impediment to forming *actual* communities—including online communities." Here Burbules' makes the point that any single conception of what makes for a good learning community is necessarily flawed because it invariably only accounts for a portion of possible learning styles. The excess will be ignored. Ihde uses a similar argument in relation to the new-networked society and globalization in general. For Ihde, while this new world poses dangers it also opens up the possibility of a genuinely global community, promoting the recognition of a plurality of cultures and challenging the Eurocentrism of the West. Ihde (1993) illustrates the point thus: "the dramatic space shots of earth from the moon or a satellite are very un-Heideggarian precisely because they place earth at a distance from earth-as-ground. But they are also irreversibly part of the post-modern view of earth-as-globe with a very different sense of what constitutes our 'home'."

Modern technology has irrevocably altered the way that we understand ourselves in relation to the world and each other. To suggest that the only "authentic" model of human relations is that of the local face-to-face encounter fails to address the reality of the globalized world and the potential for technologies to promote deeper and more complex ways of relating to others and of understanding ourselves. Notions of proximity and closeness need to be revised in accordance with the kind of proximity that the Internet makes possible. As Ihde suggests above, we now have a new sense of home, but it is home nonetheless.

CONCLUSION

While the use of the Internet within research degrees is likely to only increase, questions as to the nature of the learning that it promotes are set to become more pressing. The phenomenological insights of Heidegger, Dreyfus and Ihde cannot answer once and for all the question of whether on-line research communities and the research practices promoted by the Internet are necessarily a good or bad thing or conducive or detract from a vibrant research culture. What they can do, however, is point to the kind of questions that one might pursue in attending to these issues. In terms of understanding the impact of technology on research practices, the import of Heidegger's critique is that it focuses attention on the forms of life that technology promotes. Does the use of such technologies by researchers lead to a narrowing or broadening of practice? Do they promote one style of practice over another, randomness and eclecticism over careful consideration and focus, for example? Addressing these questions has the potential to shed light on the broader question of how the spatial re-ordering of the Internet might impact on what is researched by research students and how. In some ways, only time will tell. But a phenomenological analysis highlights the way in which technologies like the Internet cannot be understood as merely neutral "tools." Rather, such technologies actively re-orient our behaviour having far reaching consequences.

In terms of these consequences, Dreyfus appears unable to reconcile traditional notions of community and research practice with the forms of life promoted by the Internet. By contrast, for Ihde, the value of the Internet can only be determined from within the world that it opens up. Ihde's phenomenology of technology suggests that the way forward in terms of how digitally mediated research practices and communities might be understood is to engage with the uniqueness of the emergent forms of life that arise from the Internet itself. While Ihde is cautious not to romanticize the Internet he points toward its potential to displace the parochialism of the pre-global era. Looked at this way, the role of the Internet within research practices and communities, however conceived, could be to promote a model of learning more conducive to being able to deal with the problems and potentials of this globalized era within which we find ourselves.

NOTES

1. In Australia research degrees are characterised by at least two-thirds research and no more than one-third coursework
2. I insert this caveat yet again as a reminder that the politics of access are not absent from the realm of the virtual.

REFERENCES

Agre, P. (1999). The distances of education. *Academe, 85*(5), 37–41.

Barnacle, R. (2001). *On-going postgraduate research experience survey: Report and findings.* Melbourne: RMIT University.

Barnacle, R. (2002). *Investigating part time research students in professional work—A pilot study.* Melbourne: RMIT University.

Barrett, E., & Lally, V. (2000). Meeting new challenges in educational research training: The signposts for educational research cd-rom." *British Educational Research Journal, 26*(2), 271–290. Retrieved March 2002 from ProQuest Education Complete database.

Brehony, K. J., & Deem, R. (2000). Doctoral students' access to research cultures—are some more unequal than others? *Studies in Higher Education, 2*(2), 149–165. Retrieved March 2002 from ProQuest Education Complete database.

Brown, H., (et al.) (Eds.). (2001). *Politics of a digital present: An inventory of australian net culture, criticism and theory.* Melbourne: Fibreculture Publications.

Burbules, N. C. (2000a). *Does the internet constitute a global educational community?* Retrieved June 2002 from http://faculty.ed.uiuc.edu/burbules/ncb/papers/community.html.

Burbules, N. C. (2000b). *Globalization and education: An introduction.* Retrieved June 2002 from http://faculty.ed.uiuc.edu/burbules/ncb/papers/global.html.

Corry, M., & Tu, C-H. (2002). A paradigm shift for online community research. *Distance Education, 22*(2), 245–263.

Delamont, S., Atkinson, P., & Parry, O. (1997). Critical mass and doctoral research: Reflections on the Harris report. *Studies in Higher Education, 22*(3), 227, 319–331.

de Valero, Y. F. (2001). Departmental factors affecting time-to-degree and completion rates of doctoral students at one land-grant research institution. *Journal of Higher Education,* 7(3), 341–367. Retrieved March 2002 from ProQuest Education Complete database.

Dreyfus, H. (1996). *Highway bridges and feasts: Heidegger and Borgmann on how to affirm technology.* Retrieved May 2002 from http://socrates.berkeley.edu/~hdreyfus/html/paper_highway.html.

Dreyfus, H. (2001). *On the internet.* London: Routledge.

Edwards, R., & Usher, R. (1998). 'Moving' experiences: Globalization, pedagogy and experiential learning. *Studies in Continuing Education, 20*(2), 159–174.

Edwards, R., & Usher, R. (2000). *Globalization and pedagogy: Space, place and identity.* London: Routledge.

Evans, T. (2002). Part-time research students: Are they producing knowledge where it counts?, forthcoming: *Higher Education Research and Development.*

Harris, H. (1996). *Review of postgraduate education (Section 5.36).* Higher Education Funding Council for England: United Kingdom. Retrieved April 2002 from: http://www.hefce.ac.uk/pubs/hefce/1996/m14_96.htm.

Heidegger, M. (1993). The question concerning technology. In D. F. Krell. (Ed.), *Basic writings: From being and time (1927) to the task of thinking (1964).* San Francisco: HarperSanFrancisco.

Hung, D., & Chen, D-T. (2002). Understanding how thriving internet quasi-communities work: distinguishing between learning a*bout* and learning to b*e*" *Educational Technology,* January–February, 23–27.

Ihde, D. (Year of publication unknown). *Expanding hermeneutics.* Retrieved May 2002 from http://www.sunysb.edu/philosophy/faculty/papers/Expherm.htm.

Ihde, D. (1993). *Postphenomenology: Essays in the postmodern context.* Evanston, IL.: Northwestern University Press.

Kemp, D. (1999a). *New knowledge, new opportunities: A discussion paper on higher education research and research training.* Commonwealth of Australia: Canberra. Retrieved April 2002 from: http://www.dest.gov.au/archive/highered/otherpub/greenpaper/contents.htm.

Kemp, D. (1999b). *Knowledge and innovation: A policy statement on research and research training.* Commonwealth of Australia: Canberra. Retrieved April 2002 from: http://www.dest.gov.au/archive/highered/whitepaper/heading.htm.

Lovink, G. (2002). Philosophy: Review of dreyfus: On the internet. Message posted to discussion list at fibreculture@lists.myspinach.org, 19 May, or see archive at www.fibreculture.org.

Lovitts, B., & Nelson, C. (2000). The hidden crisis in graduate education: Attrition from PhD programs. *Academe, 86*(6), 44–50. Retrieved in March 2002 from ProQuest Education Complete database.

Manovich, L. (2001). *The language of new media.* Cambridge, Massachusetts: The MIT Press.

Moran, D. (2000). *Introduction to phenomenology.* London: Routledge.

Pearson, M. (1999). The changing environment for doctoral education in Australia. *Higher Education Research and Development, 18*(3), 269–285.

Pearson, M., & Ford, L. (1997). *Open and flexible PhD study and research.* Department of Education, Training and Youth Affairs: Commonwealth of Australia.

Sakamoto, T. (2002). Educational reform based on e-learning: The establishment of an international web-based learning community. *Journal of Studies in International Education, 6*(2), 156–171.

Urry, J. (1998). Contemporary transformations of time and space. In P. Scott (Ed.), *The globalization of higher education.* Buckingham, UK: SRHE and the Open University Press.

Usher, R. (2002). Putting space back on the map: Globalization, place and identity. *Educational Philosophy and Theory, 34*(1), 41–55.

CHAPTER 13

GEOGRAPHIES OF RESISTANCE IN CRITICAL PEDAGOGIC PRACTICES

Michael Peters

How can we have been misled into thinking "the social" and "the educational" without the consideration of space? The notion of space and the politics of space implicitly enter into conceptions of both "the public" and "the body" (the body politic). The public is inherently spatial, referring to the structure and the morphology of a shared place. Public spaces exist in many different forms; boulevards, parks and gardens but also cafes, malls, markets and city-centres. These spaces can also be enclosed spaces of public institutions such as universities, schools and libraries. Habermas has written of the development of the "public sphere," a developing public space that is intimately connected to the logic of democracy and citizenship, and hence, public debate and participation.

It is also important to distinguish between the exploration of space as a sociological issue—the spatial dimensions of social processes—and the theoretical use of spatial master-metaphors. Siber (1995) comments on the increasing currency of spatial metaphors—"fields," "space," "boundaries"—in contemporary sociological thinking, from Marx's infrastructure/superstructure distinction, Simmel's "social distance and proximity," Sorokin's idea of sociocultural space, to Giddens "structuration theory." She focuses, first, upon Bourdieu's influential conceptions of social space

Space, Curriculum, and Learning, pages 189–200

and field(s) as the strongest theoretical use of spatial metaphors in sociological theory. Secondly, she discusses Harrison White's (1992) *Identity and Control* as the first systematic statement of "network theory," before examining what she calls "weaker" usages of spatial metaphors and the theoretical logic of spatial metaphors. Silber (1995) concludes that spatial metaphors have begun to displace constructs and metaphors emblematic of positivism while also reflecting the prevalent distrust of any kind of encompassing and totalizing paradigm.

In this chapter I build upon my earlier work (Peters, 1996) where I outlined"architecture of resistance" in educational theory. There I examined geographical space and what I called "the post modernization of education," especially in relation to Foucault's work on institutional enclosure and the hope of a "critical regionalism," and, finally, what I called "networked spaces of societies of control." In particular, I was interested in mapping the new open spaces of education, based on processes of free-floating control, which characterized the shifts from disciplinary to control societies, from closed to open systems. I ended with a plea to take the politics of space seriously and, in particular, to investigate the way that the new spatiality was above all a questioning of historicism and historicist assumptions in educational thinking. I also emphasized that the politics of space is based upon the way in which it is fundamental to the exercise of power. We can appreciate Foucault's insight by postulating the relationships between space and power at the macro level (witness the new global political economy of education based on the open system); at the meso level (in terms of the rearrangement of institutional spaces); and at the micro level (in relation to "the classroom" or the "lecture theatre").

In this chapter I take a different tack by reviewing work that explores a Heideggerian-Foucaultian line of thinking (Elden, 2001), seeking to develop this perspective for a re-consideration of space as an essential part of critical pedagogic practices. The chapter begins from an examination of Heidegger's work to consider the importance of space in relation to the understanding of time, history and Being. Heidegger's *ontology of space* in relation to *Dasein*, or being-in-the-world, is a prominent theme in his later philosophy, one that is crucial to understanding his critique of modern technology and of modernity in general. It provides an important set of considerations not only for understanding pedagogy in modernity, but also for understanding pedagogical practice in relation to the contemporary technologization of education and, in particular, various technological forms of education, including distance education. I explore the question of spatial ontology in relation to Dreyfus' (2001) *On the Internet*, as an exemplification of the critical issues it raises for pedagogy. In the next section, the chapter investigates the Heidegger-Foucault connection and Foucault's (1986) "Of Other Spaces" as a basis for re-conceptualizing the relation between power, knowledge and the body. Foucault demonstrates the relations between these three in spatial terms and by reference to disciplinary societies based upon forms of enclosure. The body occupies space. It is an analysis of the body that must become the basis for pedagogical practices that are critical. Finally, I explore the implications of this line of argument for geographies of resistance of in critical pedagogic practices. The argument, then, is at one and the same time an explication of the importance of space and an analysis of space in terms of Heidegger's ontology, Dreyfus' phenomenology of learning, and Foucault's analysis of disciplinary societies.

HEIDEGGER AND THE SPATIAL ONTOLOGY OF BEING

The notion of "dwelling" represents Heidegger's (1971) reappraisal of the concept of space in relation to being, or as he terms it, Dasein. Young (2002) identifies three phenomena, what he calls after Heidegger, "the loss of the gods, the violence of technology, and homelessness, the loss of dwelling" as the destitution or sickness of modernity. Modern technology conquers space, abolishing "remoteness." At the same time, paradoxically, in its "abolition of remoteness," technology also brings about a profound "distancing" from things and the world. As Malpas (2000) explains:

> Although technology is, as Heidegger acknowledges, a *mode* of disclosedness or revealing, its particular "en-framing" of things also entails a covering-over of things as they extend beyond the technological frame: within the domain of the technological, things are disclosed, not as *things*, but as *resources, material,* or "*stock*" ("*Bestand*"—often translated as "standing-reserve")—as commodities to be transformed, stored, and consumed in a way that obliterates difference and renders everything in a one-dimensional sameness.

Thus modern technology is reduced to a uniform system of consumable resources and while disclosedness or revealing still occurs, it is hidden from us. Disclosedness is only possible through a certain form of being-in-the-world or "dwelling," in which things are revealed to us in their complex embeddedness in the world. The problem is that technology threatens this possibility of "dwelling." Not surprisingly, Young (2000) maintains "dwelling can plausibly be said to constitute the central topic of the thinking of the late Heidegger."

Malpas (2000) claims that spatiality plays a central role both in Heidegger's critique of modern technology (and modernity) and also in his account of being-in-the-world (dwelling). Yet in Heidegger's early work (1961) spatiality is both derived from and considered as secondary to temporality. However, Dreyfus (1992) claims that Heidegger confuses two senses of "Being-in" when he coins the term "Being-in-the-world": a spatial sense (as "in a box"—"inclusion" or "containment") and an existential sense ("in the army," "in love"—"involvement"). The spatial sense Heidegger understands in a characteristically modern sense as bodily extension. Dreyfus' (1992) criticism is that Heidegger fails to distinguish *public* space from the centered spatiality of each *individual* human being. The central difficulty therefore is that the early Heidegger derives spatiality from temporality and can only offer an impoverished sense of space. He is therefore unable to offer a consistent account of space or locatedness, which Malpas (2000) claims is directly tied to his inadequate treatment of the body.

Rather than talking of space or spatiality, Malpas (2000) suggests that we adopt the notion of *place* "as a structure that encompasses ... both spatial and temporal structures as well as structures deriving from the equipmental and the social." Extending Heidegger's line of argument, Malpas explains that within technological modernity, the spatial ordering of things is disturbed so that the difference between the near and far becomes obliterated. Technological disclosedness or revealing, thus covers over things. "It replaces the things themselves with images or representations ... of things—that is, it re-presents things within a particular 'frame' and in a way that is abstracted from their original locatedness" (Malpas, 2000). Malpas argues that the technological shift from locateness to dislocation, from things to representations, has a number of long-term and appalling consequences:

> Inasmuch as technology covers over the concrete locatedness of both human being and worldly object, it covers over the nature of both human and the thing. As a result, technology leads us to misunderstand the character of our own being-in-the-world as well as the mode of being of the things with which we are engaged. In this respect, inasmuch as technology removes us from our proper place, obliterating any proper sense of place, it also covers our own mortality—a mortality that can itself be viewed as essentially connected to the place-bound character of our being. (Malpas, 2000)

While seeming to be amenable to human control, modern technology therefore increasingly enframes us as resources, transforming the character of human experience, reducing it to re-presentation and narrowing our sensitivity to a small range of sensory and interactive modalities, rather than engaging the body and its multiple senses in its entirety. These Heideggerian observations are of overwhelming importance to pedagogy and to critical pedagogical practices. They provide the basis for a philosophical account of the critical importance of space and the body to pedagogy and an implicit critique of a traditional pedagogy that has all but ignored the body (space) or privileged it over mind (atemporality). They also contain within them a pedagogical critique of a technological modernity aimed at the efficient delivery of information via a model of distance education. Let me examine this line of argument further by reference to the work of Hubert Dreyfus.

BODY TALK AND NIHILISM

Dreyfus' (2001) *On the Internet* is at one and the same time, philosophical, post-Nietzschean and also 'computer literate' or, better, computer sensitive. His (2001) thesis on the Internet is succinctly summed up in the final paragraphs:

> as long as we continue to affirm our bodies, the Net can be useful to us in spite of its tendency to offer the worst of a series of asymmetric trade-offs: economy over efficiency in education, the virtual over the real in our relation to things and people, and anonymity over commitment in our lives. But, in using it, we have to remember that our culture has already fallen twice for the Platonic/Christian temptation to try to get rid of our vulnerable bodies, and has ended in nihilism. This time around we must resist this temptation and affirm our bodies, not in spite of their finitude and vulnerability, but because, without our bodies, as Nietzsche saw, we would be literally nothing.

It is a thesis as powerful as it is frightening, as simple and elegant as it is prophetic. The Net as a kind of technological *enframing* of being stands at the door. It contains both the danger and the saving power. If we allow it to transcend the limits of the body, we will also allow it to abstract from our moods, our cultural location and belongingness, our finitude and vulnerability, our animality that helps comprise our linguistic and cultural identities, and also the meaning we give our lives. By leaving the body behind we will succumb to the same nihilistic impulses in our culture that began with Platonism and was repeated by Christianity (Peters, 2002).

Platonism in educational philosophy stands for the elevation and privileging of the mind or intellect over the body and optional metaphors that serve to bifurcate reason and emotion. In their application and formalization such metaphors have become the substance of educational practice. Perhaps, the most culturally deeply embedded dualism with which educational theory and practice must come to terms with is the mind/body separation. This dualism historically has developed as an instrument of "othering," of separating boys from girls, reason from emotion, minorities from the dominant culture, and classes from each other. It nests within a family of related dualisms and remains one of the most trenchant and resistant problems of education in postmodernity.

The same underlying philosophical problem-set is at issue in Dreyfus' work. The body is everything. Dreyfus (2001) argues "loss of embodiment would lead to *loss of the ability to recognize relevance*" and, most importantly for learning, that "without involvement and presence *we cannot acquire skills.*" In short, our bodies matter. They provide the "source of our sense of our grip on reality" (ibid.). He is a long-standing critic of computer ideologies and, in particular, the cognitive science modelling of the brain on the computer. In *Mind over Machine* (Dreyfus & Dreyfus, 1986) he argued that where human beings begin learning a new skill by understanding and carrying out its rules as a novice, only to leave rules behind as they become expert, the best that a computer can attain is a sort of "competence" that consists in carrying out the rules it has been taught, albeit very quickly and reliably. A version of this argument appears in *On the Internet*. Indeed, Dreyfus outlines the stages by which a student learns—novice, advanced beginner, competence, proficiency, expertise, and practical wisdom. Dreyfus major conclusion is that disembodied learning can only ever attain the stage of mere competence and will not achieve the stage of self-mastery or practical wisdom. Questions of space and the body, of the location and place of the learner and teacher, are, thus, crucially germane to the phenomenology of learning, as they determine questions of *locatedness*, *situatedness*, and, therefore, also broader notions of identity, at the personal, national and cultural levels.

FOUCAULT'S SPACES OF POWER

The analysis of space was always important to Foucault, although he wrote little directly on this topic. Foucault (1986) held that space is inherently political and fundamental to any exercise of power. Under the influence of Heidegger and the movement of structuralism more generally, Foucault provided a history of the present that recast genealogy as a historical ontology, a form of *spatialized history*, rather than merely a history of space (Elden, 2001). An equally important aspect of Foucault's analysis of space is his emphasis on the materiality of the body.

For Foucault, every social space is structured by power and invested with knowledge. In a late interview he remarked that in the eighteenth century there emerged a specific political discourse, which reflects upon architecture as an important aspect of the art of government. Cities and buildings became spatial models for government rationality. They provided the basis for the exercise of social control and manipulation. The school, for example, is considered as a form of disciplinary architecture demonstrating at an abstract level the relation between educational space and a particular form of disciplinary political rationality that produces an *individualized* subject.

Foucault painstakingly documents the histories of different institutional spaces: the clinic, the prison, and the school. He observes how "disciplinary power" depends upon "a politics of space."

> A whole problematic then develops: that of an architecture that is no longer built simply to be seen (as with the ostentation of palaces), or to observe the external space (cf. the geometry of fortresses), but to permit an internal, articulated and detailed control—to render visible those who are inside it; in more general terms, an architecture that would operate to transform individuals: to act on those it shelters, to provide a hold on their conduct, to carry the effects of power right to them, to make it possible to know them, to alter them. (Foucault 1979)

Based on the instruments of observation, judgement and examination, disciplinary power is enabled by a carefully designed institutional architecture permitting total surveillance. The hospital building is organized as "an instrument of medical action." The prison is built as a "space of confinement." The school building is organized as a spatial mechanism for the training of the individual, for individualization. He calls it "a pedagogical machine" (ibid.). The result is discipline. During the seventeenth and eighteenth centuries the disciplines became general formulas of domination. Foucault claims that this new political anatomy was evidenced in a multiplicity of often-minor processes at different locations that eventually coalesced into a general method. "They [i.e., disciplinary techniques] were at work in secondary education at a very early date, later in primary schools; they slowly invested the space of the hospital; and, in a few decades, they restructured the military organization." Foucault talks of disciplinary techniques in terms of "the art of distributions," (the monastic model of enclosure became the most perfect educational regime) and "partitioning" (every individual had his or her own place). "The rule of *functional sites*" refers to the ways that architects designed space to correspond to the need to supervise and to prevent "dangerous communication." Foucault (1979) argues "the organization of a serial space was one of the great technical mutations of elementary education" that made it possible to supersede the traditional apprenticeship system where the pupil spends a few minutes with the master while the rest of the group remains idle.

> To sum up, it might be said that discipline creates out of the bodies it controls four types of individuality, or rather an individuality that is endowed with four characteristics; it is cellular (by play of spatial distribution), it is organic (by the coding of activities), it is genetic (by the accumulation of time), it combinatory (by the composition of forces). And, in doing so, it operates four great techniques; it draws up tables; it prescribes movements; it

> imposes exercises; lastly, in order to obtain the combination of forces, it arranges "tactics." (Foucault, 1979)

He discusses the means of correct training in terms of "hierarchical observation." As he suggests "the school building was to be a mechanism for training ... a pedagogical machine" (ibid.) for normalizing judgement. The examination "transformed the economy of visibility into the exercise of power," introduced "individuality into the field of documentation," and "surrounded by all its documentary techniques, ...[made] each individual a 'case'" (Foucault, 1979).

Dreyfus' (1996) contrasts the work of Heidegger and Foucault thus:

> At the heart of Heidegger's thought is the notion of being, and the same could be said of power in the works of Foucault. The history of being gives Heidegger a perspective from which to understand how in our modern world *things* have been turned into *objects*. Foucault transforms Heidegger's focus on *things* to a focus on *selves* and how they became *subjects*. And, just as Heidegger offers a history of being, culminating in the technological understanding of being, in order to help us understand and overcome our current way of dealing with things as objects and resources, Foucault analyses several regimes of power, culminating in modern bio-power, in order to help us free ourselves from understanding ourselves as subjects.

In contrast to Platonism and in line with the trajectory of contemporary French philosophy Foucault emphasizes the enfleshed subject, sensuous reason, and the embodied subject. This is a subject, not timelessly cast as an abstract universalism as with the *homo economicus* of neoclassical economics, but one that emphasizes the everyday contingencies of the self and its understanding as the site of a dissociated self—a body normalised, individualised and inscribed with the effects of power. While Dreyfus does not provide a point of comparison between Heidegger and Foucault specifically in relation to space, he does provide a useful analysis of their respective notions of resistance in relation to technology and bio-power.

> Neither Heidegger nor Foucault think that we can resist techno/bio-power directly because what ultimately needs to be resisted is not particular technologies or particular strategies but rather a tendency in the practices towards ever greater order and flexibility that produces and sustains them. Thus the current understanding can only be resisted by first showing that it is not inevitable but is an interpretation of what it is to be, second by connecting our current style with our current discomfort and then by taking up marginal practices which have escaped or successfully resisted the spread of techno/bio-power.

Foucault bases his notion of resistance on the self. His genealogical investigations and historical ontology leads him back to the ethics of self-mastery, a set of cultural practices of the self he finds in the writings of the Stoics that provide a basis to demonstrate that there are no universal necessities in human existence (that we could be otherwise). In his last seminars, given at Vermont and Berkeley in 1983, Foucault illustrates by reference to classical texts the form of an education, differently conceived at different times, based on truth-telling and a conception of self-mastery that integrates the *logos* with the *nomos* and the *bios* in practical ways that lead to a successful engagement with the world (Peters, 2003). In terms of these variable practices of the self he is able to question the direction of our current practices, and to contemplate the possibility of resistance. In particular, he draws upon the question in Antiquity of knowing how to govern one's life so as to give it the most beautiful form possible. Here the notions of the "art of living," "care of the self" and "aesthetics of existence" loom large and Heidegger's and Nietzsche's combined influence is keenly felt, especially in the notion of life as a work of art. Ultimately it is in these "practices of creativity" of the self that Foucault wants to ground the possibilities of resistance. Dreyfus (1996) comments:

> There is, nonetheless, an important kind of resistance these two thinkers [Heidegger and Foucault] share. Thinking the history of being, for Heidegger, and the genealogy of regimes of power, for Foucault, opens a space for critical questioning by showing that our understanding of reality need not be defined by techno/bio power—that we need not be dominated by the drive to order and optimize everything. They both show that we had a different relation to being and to power once, which suggests that we could have again. Thus an understanding of our historical condition weakens the hold our current understanding has on us and makes possible disengagement from the direction our practices are taking.

In one sense it could be argued that resistance for both Heidegger and Foucault is retrospective and based on a form of historical understanding that demonstrates that we could be otherwise than we are. Yet there are positive strategies and tactics of resistance as well to be gained from their work, which are important to mapping geographies of resistance in critical pedagogical practices. These forms of resistance crucially involve the body and the relations between the body and place, which also provides the ground for most discussions of identity, citizenship and the other. Yet if we are to understand resistance in Foucaultian terms, we must also understand that his conception of power *entails* resistance—resistance is a relation of power. Just as freedom is a necessary condition for relations of power, so it is for resistance. Foucault does not accept the liberal juridical notion of power, which is grounded in sovereignty and opposed to authority. He understands power (and resistance) as something to be exercised rather than something to be possessed. Power is productive (as well as repressive) and it is diffuse, permeating the social body through multiple sites. On this model, then, where the very existence of power depends on points of resistance, critical pedagogical practices depend upon on an embodied engagement with the world, based on the ethic of self-mastery that seeks to "recover" our own sense of place and the potential of modern technology to obliterate all difference.

GEOGRAPHIES OF RESISTANCE IN CRITICAL PEDAGOGIC PRACTICES

In a recent essay Morgan (2000) has argued for a "critical pedagogy of space" where space is seen as a social construction. He (ibid.) suggests that "space should not be seen simply as the product of capitalist social relationships, but is tied up with other axes of power, such as gender, ethnicity and sexuality" and he ends by issuing the challenge to develop a critical pedagogy of space that reflects the multiple and contested nature of space. He emphasizes the way in which the production of space is linked with the exercise of economic power, what in the context of globalization I have called the "post modernization of education." He then considers space in relation to the school curriculum, especially geography, where he indicates "different social groups may have distinct spatialities" (ibid.), emphasizing "a shift away from the idea of space as homogenuous, continuous, objective, Cartesian and knowable, towards a view of space as fragmented, imaginative, unknowable and subjective" (ibid.). He raises the possibility that "a critical pedagogy of space might begin with an analysis of the gendered use of space in the classroom" (ibid.) and it might also pay attention to "the experience of public space," focusing on "spaces of exclusion," and "geographies of resistance through which people deal with, and resist, oppressive practices" (ibid.). He concludes by suggesting that "spaces as social texts" provide the opportunity for highlighting issues concerning identity formation in relation to questions of scale "through the *body-home-community-city-region-national-global*" (ibid.).

Morgan's account is a useful addition to the literature on space, although it is restricted largely to the ways in which critical pedagogy might be encouraged in academic geography. It also depends in large measure on"geography of resistance" in the narrow disciplinary sense—how we might import new politicised notions of space into school geography. This is undoubtedly an important project but the notion of "geographies of resistance" needs to be interpreted more broadly and not restricted to one subject in the curriculum where it has obvious application. Clearly, the themes *body-home-community-city-region-national-global* may be easily pursued in other parts of the arts/humanities curriculum.

More importantly, there is a need for theorists and practitioners in critical pedagogy to address questions concerning critical *ontologies of space* by unpacking the notion of the body, construed both in terms of the historically privileged set of mind/body and its impact on learning theory (i.e., in terms of the individual learner) and in term of collective notions (the student body). Here the contributions of Heidegger and Foucault are of paramount importance, not simply because they provide one trajectory for a consideration of the importance of space philosophically—from an essentializing ontology of being (Heidegger) to an historical ontology of ourselves (Foucault)—but also because Foucault's opus depends crucially upon spatial concepts that he has developed in relation to institutional spaces of enclosure, to *epistemes* (and the spatalisation of knowledge and their disciplinary formation), and to the transition from disciplinary societies to societies of control (Peters, 2001). Geographies of resistance are geographies of the body that are intimately tied to the politics of space, to locality and to identity and identity politics. Critical pedagogical practices considered in this sense may not, therefore, be related only to contents areas and their successful prosecution in the classroom, but may indeed more properly embrace lived practices and conditions of existence.

REFERENCES

Dreyfus, H. (1992). *What computers still can't do: A critique of artificial reason.* Cambridge, MA: MIT Press.

Dreyfus, H. (1996). Being and power: Heidegger and Foucault. Originally published as "On the ordering of things: Being and power in Heidegger and Foucault, *Michel Foucault, Philosophe,* Le Seuil, Paris, (1989). Reprinted in *International Journal of Philosophical Studies,* Vol. 4. Also available at: (http://socrates.berkeley.edu/~hdreyfus/html/paper_being.html).

Dreyfus, H. (2001) *On the internet.* London and New York: Routledge.

Dreyfus, H., Dreyfus, S., with T. Athanasiou. (1986). *Mind over machine: The power of human intuition and expertise in the era of the computer.* New York: Free Press.

Elden, S. (2001). *Mapping thepPresent: Heidegger, Foucault and the project of a spatial history.* London and New York: Continuum.

Foucault, M. (1979). *Discipline and punish: The birth of the prison.* London: Penguin.

Foucault, M. (1986). Of other spaces. *Diacritics, 16*(1).

Heidegger, M. (1962). *Being and time.* London: SCM Press.

Heidegger, M. (1971). *Poetry, language, thought.* New York London: Harper and Row.

Malpas, J. (2000). Uncovering the space of disclosedness: Heidegger, technology, and the problem of spatiality in being and time. In M. Wrathall & J. Malpass (Eds.), *Heidegger, authenticity, and modernity: Essays in honor of Hubert L. Dreyfus, Vol 1.* pp. 205–227. Cambridge, MA.: The MIT Press.

Morgan, J. (2000). Critical pedagogy: The spaces that make the difference. *Pedagogy, Culture and Society, 8*(3), 273–289.

Peters, M. A. (1996). *Poststructuralism, politics and education.* Westport, CT and London: Bergin and Garvey.

Peters, M. A. (2001). *Poststructuralism, marxism and neoliberalism: Between theory and politics.* Boulder, NY: Rowman and Littlefield.

Peters, M.A. (2002). Dreyfus on the internet: Platonism, body talk and nihilism. *Educational Philosophy and Theory, 34*(4).

Peters, M.A. (2003). Truth-telling as an educational practice of the self: Foucault, p*arrhesia* and the ethics of subjectivity. *Oxford Review of Education,* (forthcoming).

Silber, I. F. (1995). Space, fields, boundaries: The rise of spatial metaphors in contemporary sociological theory. *Social Research, 62*(2), 323–333.

White, H. C. (1992). *Identity and control: A structural theory of social action.* Princeton, NJ: Princeton University Press.

Young, J. (2002). *Heidegger's later philosophy.* Cambridge: Cambridge University Press.

CHAPTER 14

ABSTRACTED SPACE

The World, The Net, and The Contemporary University

Robin Usher

The title of this chapter suggests that what follows could verge on the inflated and perhaps even the pretentious. There is an implication that everything will be definitively explained. However, another reading is possible, an ironical one where the title is deliberately presented to provoke, since definitive explanations are problematic and dangerous. Thus in this chapter I am not attempting to prove anything, either conceptually or empirically. What I try to do is put forward an argument, which could turn out to be interesting and provocative. It is however not a totalizing argument, perhaps not even an entirely consistent one and it is certainly a partial one. Unlike the implication of the title, I am struck by the fact that since the contemporary moment is ephemeral and highly accelerated, no one can really know what any social reality will look like tomorrow let alone in the future beyond (http://pscw.unva.nl/SOCIOSITE/websoc/introE.html).

If pressed, I would say this chapter is about space and technology. Not technology in the sense of hardware but technology as *mediation* and the mediation that transforms that which it mediates. I deploy this notion of technology to consider the space of the university, in particular the intel-

Space, Curriculum, and Learning, pages 201–220

lectual practices embedded in that space. I argue that this space is both a networked space and a space of *abstraction* and that these are intimately interconnected. I also argue that new information and communication technologies (ICTs) have been crucial in enabling the growth of the techno sciences, the information society and the knowledge economy and in valorizing the mode of performativity associated with these developments. In turn, these have affected the space of the contemporary university to the extent that its form of life mirrors that of the information society and is likewise mirrored by the latter. This leads me to consider the nature of this abstracted space in relation to the immediacy, accessibility and increasing ubiquity of the Net (http://pscw.uva.n1/SOCIOSITE/websoc/introE.html). I argue that abstracted space in these conditions maps a highly extended interconnectivity, perhaps even an unlimited connectivity, and a network of connections that are extended globally and timelessly. This then raises the issue of the virtual, as another expression of abstraction, in relation to the face-to-face, or to put it another way, how necessary are mutual relations of presence, in the production, valorization and use of knowledge?

In developing these arguments in this chapter I am conscious that I too am engaging in an intellectual practice that both occupies and constructs a space of abstraction and which itself involves a process of technological mediation. In constructing this space, I locate myself in the spaces of intellectual life by articulating and exemplifying my argument in and through the technology not only of writing but also of electronic text production. My location is grounded in intellectual practices whose form is that of "standing back" and the following of certain tacit rules of interaction that enable the expression of this abstraction in an acceptable form. My practice is one of writing in academic prose where abstract, in the sense of generalized, arguments and addressed to an anonymous audience are expressed through a linear and coherent text. At the same time, however, I recognize that in writing this text I am immersed in the virtual and indeed this text would not have been possible without that immersion. Consequently, I have tried to self-referentially recognize this through a mild subversion of the conventions of the academic text. I retain a location in the space of abstraction but also bring to the fore the virtual by drawing on the resources of the Net and embedding the references to these in the "body" of my text. At one level this can be understood as just the conventional practice of attributing one's sources. At another level I am reinforcing something said by Kellner about his intellectual practices: "The major research for two of my recent books was done from computer databases and I find that I am spending much more time in databases than in libraries for a variety of research projects" (http://www.gseis.ucla.edu/courses/ed235a/dk/INT.htm).

At yet another level however I am also trying to exemplify the impact and productivity of the Net and the WWW, of how there are now unlimited transnational global knowledge webs. Of course, I could have done so more effectively if I had not had to produce a text for a printed book but rather had been able to post this text on to a Website with hyperlinked references. No doubt however this at least exemplifies the continuing power of the printed text as a mediating technology.

SOCRATES IS NOW ONLINE

I am writing this chapter sitting at my laptop and as I do so I reflect on the scene of writing and the shift "from paper and pen or typewriter to the globally networked computer" (Poster, 2001, p. 16). I am struck by what it now means to "write" and how the computer has transformed the meaning and practice of writing academic texts. In this respect, I agree with Bolter's (1991) comment that "the computer is the ideal writing space for our networked society" (http://www.gse.uci.edu/markw/info-age.html).

At this point, a fragment of autobiography will make this clearer. I am not of the generation that was brought up with a computer immediately to hand. Computers came relatively late into my personal and professional life and I can still vividly remember the time when one wrote with pen and paper rather than pressing keys on a computer. When I reflect on this I realize that not only is this a lost age never to return but it is also inconceivable that we could ever produce and manage knowledge in the same way; as Hasan (2001) puts it, "the PC, email and—increasingly—the Internet have become essential tools for academics and students alike." In this sense, intellectual practices that must inevitably involve writing have indeed changed. However whilst it is clear that for many others and myself the computer, and information technology generally, is an indispensable prosthesis, is there more to it than this? Has computer-mediated writing not only made writing less laborious but also changed what the intellectual practice of knowledge production involves and implies, particularly in terms of its relational spaces. This leads to the wider question of whether the Net should be seen in purely instrumental terms, as a handy tool, which Poster argues, is problematic since it hinders understanding of the Net's ever changing form and the potential that such a change might have: "as long as we remain within an instrumental framework we cannot question it, defines its limits or look to new media in relation to how it might generate new cultures" (Poster, 2001, pp. 91–92).

The Net is still largely a medium of the written text, although this is changing rapidly with the growth of multimedia. At the present time, however, the production of knowledge texts through writing is still very much with us. Despite the profound impact of ICTs in other ways, it is probably very likely that writing will continue to be the basic medium of online expression, the skeleton around which other technologies and experiences will continue to be organized (http://commons.somewhere.com/rre/2002/RRE.Critical.Theory.of.T.html). The computer has certainly enabled the generation of a great deal more writing, although there is a furious debate about whether this constitutes knowledge or 'mere' information. Also, to the extent that hypertext is becoming more common as a medium for the writing of text, then the way we write is also changing. This raises the question about the social form or mode of relationality that is produced in this space.

In this chapter I want to present a position that brings to the fore technological mediation by looking at intellectual practices in relation to ICTs. Writing as a mediating technology, between the spoken word and the written text, between thought and expression, is the premier means by which knowledge has been articulated in intellectual practices. Sharp (2002, p. 267) puts it this way: "technical mediations have been inseparable from intellectual work. The tie to writing is so especially buried in time, so radically entrenched in custom and taken for granted, that we scarcely recognise that it is a technological mediation, one which allows the transformation of speech."

Despite this "taken for grantedness," writing has nonetheless been criticized precisely because it is a technology that mediates presence. Plato believed that writing destroys the face-to-face, dialogic relationship connecting teacher and student. He argued that people will become shallow, unable to create anything original, a burden to society, have access to the wrong kinds of information and above all will become unable to interact with others (http://www.anu.edu.au/english/jems/lb/academicpapers/IPEG.html). A very similar critique has been mounted against the computer, particularly in relation to online learning. However, it is interesting to note that just as Plato used a written text to critique writing, many of the most vociferous attacks on web-based media circulate on the Net. Yes, Socrates is now very much online (http://www.stemnet.nf.ca/~elmurphy/emurphy/cle.html).

My purpose in focussing on writing as technological mediation is to make the apparently obvious point that it is a medium of communication and not just a handy tool. As such, issues to do with forms of connection and interaction come to the fore. It raises the question of whether the medium in which we communicate determines the nature and quality of our interactions. Burbules (2000, p. 346) argues that "in other spaces or places the characteristics of the environment are to some extent independent of the means used to represent them; but with the Internet these two levels are utterly intertwined Paths of movement are also connections of meaning-making." The standard argument here is that the social impact of any technology depends on how it is designed and used. Another answer, not necessarily incompatible, is that technologies are not mere means to ends but also shape worlds. The message is transformed by the technological medium that delivers it. Certainly, it would not be extreme to argue that this is happening with ICTs when one person now has the power to reach another person or a million equally easily (http://www.memex.org/meme2-03.html). This phenomenon of unlimited connectivity is shaping the contemporary world, both physically and in terms of subjectivities.

My initial argument then at this stage is that ICTs are more than just a handy and neutral tool. On the contrary, a strong argument can be made that they increasingly shape many significant dimensions of our forms of life—which in a small way is the point I was trying to make earlier in contrasting writing with pen and paper and writing with a computer as radically different forms of life.

INTELLECTUAL PRACTICES

The most productive way of understanding the relationship between technology and practices or forms of life is to move away from cause and effect explanations to ones that bring to the fore mutual interaction. Technologies have transformative powers but they themselves are shaped by sociocultural influences. This is what I assume in what follows. I now turn to my main theme which is the space occupied by the university in the contemporary moment. In so doing, I will look at the relationship between the intellectual practices of the university and the space of the social that it constructs and in which it is located. In this discussion, the issue of technology, or more specifically ICTs, will play a significant role. There is also little disagreement that the university as an institution is going through profound changes, what some would argue to be a "crisis."

I am deliberately putting the spotlight on intellectual practices because I want to highlight that these are *practices* and are therefore heterogeneously social and material. To foreground practices is to highlight what practitioners do and not what they think in the privacy of their minds. Knowledge and knowledge production always takes material forms and involves a variety of interconnected actors. Taking this position enables a focus on form rather than content, on structure rather than substance. What then is the form of these practices and what is their relationship with ICTs? My argument is that intellectual practices are both socially constructed and socially constructive; they are expressed in and through a particular form; this form defines a space within which intellectual practices take place; they are practices that are technologically mediated and their relationship to the wider social structure is changing.

Hinkson (2002) asks the question why is it that the university which traditionally has been in but not of the world is now very firmly both in *and* of it. The university both in its medieval and its modern form occupied an unworldly space. This has changed to the point where economic considerations and market values are major determinants of what the university stands for and how it conducts its activities. Hinkson goes on to make the somewhat startling claim that this situation is one where the social structure is being transformed *on the university's terms*. I say startling because this theorization on the face of it seems both counter-intuitive and counter-factual. We are more accustomed to arguments that attribute the transformation of the social structure to globalization or to the impact of ICTs with the university dragged kicking and screaming helplessly in the wake of these forces. However again it is important to remember that the reference here is to the *form* of intellectual practice rather than the substance of the university's position.

Hinkson's case is that in order to understand the university it is necessary to understand the social form taken by its intellectual practices, since it is this that structures the university. His argument is that the dominant form of intellectual practice historically has been that of extended interconnectivity. Whilst substantively it is carried out by specific persons in specific places, it possesses in its *form* a quality of abstraction which makes it seem disembedded and disembodied; "the work of the intellect takes apart bodily presence in social relations" (Hinkson, 2002, p. 262). This situation structures social exchanges in all disciplinary segments of intellectual practice. We can say therefore that intellectual practice is an abstracted space that constructs a space of abstraction. In this space, social exchange takes the form of *interconnectivity* rather than face-to-face relationality, an interconnectivity enabled by the technological mediation of writing

It is this technological mediation that makes possible the "standing back" both from persons and contexts that is the most obvious feature of this abstracted form. Historically the written text (the manuscript in the time of the medieval university, the printed book and journal in the time of the modern university) has been the main mode of technological mediation. Whilst the printed text is probably still dominant, online texts and electronic publishing are fast displacing it.

Whilst this part of the argument is reasonably uncontentious, it is not easy to see how the next step can be taken. How does this technologically mediated and abstract form of life that characterizes intellectual practices equate with the social structure generally? How is it that life world practices are becoming more like intellectual practices? Here Hinkson draws heavily on Lyotard (1984). He argues that Lyotard was right in the prescient way he predicted the coming social order of the information society. Lyotard took the view that society generally, mirroring the information order, becomes structured as a system of signs where social relations are extended, free ranging, in constant process and above all reflexive. A society of signs (Edwards & Usher, 1999) is one where an increasing range of electronic technologies multiplies a multiplicity of signs. Social relations and the materiality of the world become mediated through symbolic exchanges, through the production, circulation and reception of signifying practices where signs have value in their own right. Electronic communication and information technologies produce a world of infinitely extended flows that envelop the fabric of the life world. The consequence is, as Castells (1996) argues, that contemporary society becomes characterized by timeless time and spaceless space (http://www.georgetown.edu/crossroads/conversations/rowe.html).

This reconstruction of social life as a system of signs is an effect of the information revolution. Its impact remakes the social order. Social life is reconstructed as infinitely extended interconnection diluting the experience of presence. Experience becomes information where information has no clear connection with a subject or a concrete referent and where it flows freely and promiscuously. The more abstract social life becomes, as the system of signs takes hold, the less social life requires face-to-face mutual presence and interaction. In effect, therefore the argument is that the form of contemporary social life is increasingly that of the "virtual" rather than of presence, where the face-to-face as the social form of relationality becomes less significant.

The next question however is to do with the mechanism enabling this to happen. Hinkson's answer is that the university reproduces the information order. Here a major role is played by the techno sciences, with the contemporary university in the forefront of knowledge production in this field where IT, natural science and engineering intersect and where the emphasis is on innovation for commercialization. For Hinkson (2002, p. 260), the techno sciences are essentially about taking "nature apart and reassembling it with definite purposes in mind," rather than with investigating and interpreting in the spirit of curiosity. With the techno sciences there is no longer a distinction between the pure and the applied, the disinterested pursuit of knowledge as against its production for instrumental purposes. As they take hold, they mark a transition from an interpretive mode of being to an instrumental one where through digitization and commercialization, the emphasis is on knowledge production for market advantage, what Lyotard referred to as "performativity," that is, knowledge in the service of economic performance and efficiency. This meets a ready response in the contemporary knowledge economy where growth and competitiveness depend on the creation and deployment of intellectual capital produced by the techno sciences (http://www.j-bradford-delong.net/OpEd/virtual/technet_outline.html).

In the information society communication and computerization are significant dimensions of the knowledge complex that powers the contemporary post-industrial economy (Burbules, 2000). Digitization and the commercialization of knowledge production are inevitably linked where digitization consists of the electronic assembling, disassembling and transmitting of the basic elements of intellectual capital and where commercialization is the vehicle. Biotechnology is a good case in point of this interlinking of knowledge production, digitization and commercialization. As Jordan (1999) points out, "the basic substance of life is now understood to be made of the information code of DNA." Biology has become digitalized information that can then be commercialized by drug companies (Slaughter & Leslie, 1997).

Universities have assets of immense economic value. Knowledge embedded in intellectual capital is intimately linked to the development of technological capability. At the same time universities cannot ignore digitization since it is increasingly necessary both substantively as knowledge and as the means to produce knowledge. It provides new opportunities to extend knowledge producing capacity and to broaden alternative revenue streams. As a consequence, "through IT, the higher education sector has achieved an almost seamless convergence with the business sector" (Hasan, 2001). The university becomes porous, its currency becoming more accessible as it is increasingly drawn into the world of business (Bowen, 2000).

Whilst therefore the university in terms of the substance of knowledge becomes increasingly dependent on digitization and commercialization, the form of intellectual practices continues to be the extended interconnectivity of abstract exchange. Hinkson (2002) argues that even with the academic entrepreneur the form supporting this trend still involves "standing back" from the everyday social structure. On the face of it, this seems to contradict the notion of the university being now *of* the world as well as in it. However the dominance of the transformative techno sciences and the instrumental mode of knowledge production need to be taken into account at this point. Whilst there is still a "standing back," there is now a closer relationship or interlinking with the social structure. Hinkson (2002, p. 262) puts it this way: "the university that is of the world achieves this status by virtue of the capacity of intellectual practice to instrumentally remake the world on its own terms—the terms of a distinctive form of the social."

The virtuality and hence abstraction of the social structure now mirrors the virtual and abstracted form of intellectual practices. But why has abstraction as a form of life persisted at a time when the interpretive mode of interacting with the world is being rapidly replaced by the instrumental? Technological mediation structures the form of intellectual practice in both interpretive and instrumental modes. Moreover, as we have seen it is technological mediation in the form of globalized media, the Net and the accompanying decline of structured relations of presence that is now mirrored in the forms of life of society generally. The life world has become technologically textured and increasingly virtual. Technologically mediated social relations don't require the immediate presence of others. ICTs connect but also distance. As I have already noted, experience becomes information circulating in flows within electronic networks without a subject and therefore without the need for structured relations of presence. Social life takes the form of abstract extended interconnection, mirroring the social mode of intellectual practice. "Hyperintelligent communication offers unprecedented opportunities for people to interact across space and time but paradoxically it also distances those it links. No longer are the individuals 'commanding presences' for each other; they have become disposable experiences" (Borgmann, 1992, p. 102).

The information revolution has remade the social world in the image of social exchange typical of intellectual practice as it is now embodied in high technology, abstract, virtual and instrumental. Thus Lyotard was right when he argued that the information revolution shapes how people live with each other. As Sharp (2002, p. 283) puts it, "a fusion of the techno sciences with capital creates an unprecedented expansion of the life forms of abstracted practice…it sets in motion a radical reconstruction of all social life."

THE NET

Technological mediation of any kind creates webs of interconnectivity. Sharp (2002) refers to these webs as an "archive," a complex technologically extended and anonymous social network. The written texts of intellectual practice are detached from their authors. Knowledge production and dissemination without face-to-face interaction. Computer databases, online resources, Internet and WWW surfing, possessing characteristics of complexity, technological extension and anonymity can also be regarded as archives. "The process of virtualization links well with that of the development of a more global learning space and provides for the transition of research techniques and models based upon physical static collections to those that are virtual and easily distributable" (http://is.gseis.ucla.edu/impact/s94/students/timothy/timothy_final.html). It is unlikely that any in academic life would deny that the Net provides tremendous resources for the furthering of intellectual practices. However, the critical issue raised here is what the spatial consequences might be of this extension of the web of interconnectivity by ICTs..

The technophiles and cyber utopians are right when they highlight the significance of computers not just as a calculating and information storage device but as a means of communication which has transformative effects on social life. The Net constitutes a change in the material basis of communication and, with that, a reconfiguration of socio-cultural objects. The networks of the Net connect people through the space-time compression they make possible. They reproduce and disseminate almost without cost and largely effortlessly any type of information that people are capable of producing (http://www.computerlearning.org/articles/HypeHope.htm).

At this point I want to say something about the growth of the Net and the Web because the nature and extent of this growth is significant for my argument. At the end of 1994, 10,000 Website servers were connected to the Internet. By 1999 there were seven million. At the end of 2002, the indexable Web contains 320 million pages (http://www.neci.nec.com/~lawrence/websize.html). There are 30–40 million users of the Net worldwide. There are about 650,000 websites and 9.5 million hosts. The rate of the web's growth has been and continues to be exponential. The increase in traffic on the Internet is 15% per month. The doubling period is under six months (http://www.mit.edu:8001/people/mkgray/net/internet-growth-summary.html). Of course, it could be argued that these trends indicate only that ICTs are producing vast amounts of information. For some, this plenitude of information is simply piling trivia upon trivia. My argument however is that this very plenitude taken in conjunction with the evolving structure of the Net is stimulating changes in the economy, society and everyday life.

In the world of business, for example, even though computers have been used since the 1940s, the difference now is that the Internet raises the stakes. Computer networks have decreased transaction and co-ordination costs. By networking businesses can rationalize and standardize their activities and gain a competitive advantage. Every major business has a Website, using it to sell goods, automate its supply function and generally operate the business. Software systems especially data base management programs and enterprise communication tools now form the backbone of businesses (http://profs.lp.findlaw.com/e-commerce1.html). The economy is one where for businesses their strategic core activities play out on a global space in real time. Today's massive movements of capital depend on the communication knowledge intensive information across global markets. The bases of globalization are information and innovation both of which are highly knowledge intensive (Carnoy, 2000). In educational settings we have witnessed the development in an exponential way of greater connectivity and networking as faster networks lead to newer services and from thence to new uses. Electronic networking has penetrated into all academic disciplines, in the process expanding the space of learning environments. Virtualization is creating a global learning space—the "global matrices of minds" (http://is.gseis.ucla.edu/impact/s94/students/timothy/timothy_final.html). The irony of this is that computers do not possess an inherent logic that impelled them necessarily to become such a significant medium for communicating and producing knowledge. Socio-cultural as well as technological processes have shaped their direction of development. ICTs are reconfiguring space-time relations radically restructuring the materiality and spatiality of space and the relationship between space and place. Cyberspace has infinitely extended social interaction and through that extension changed the quality of the interaction (http://www.mappingcyberspace.com).

With the Net, there is a seeming absence of time and place. Regardless of physical place one can be in contact with much of the world via the global computer network. Online, one can live next door to every one else at any time—virtually. Geographical place has been replaced by existence in the same virtual space. The immediacy of the Net means that time is no longer a constraint on communication. Place and time are no longer the boundaries they have been. Boundaries have become more permeable (http://wuarchive.wustly.edu/doc/misc/can/netbook/ch7_Netizen).

Perhaps most significantly the world is reconceptualized as an electronic space, a space of infinite flows. The Net has made interconnectivity possible on a global scale, in the process becoming hyper-connectivity. Here, hyper-connectivity refers not only to connections in a quantitative sense but also to a situation where it is impossible to envisage the world as not being always already connected, constituted through connections and through connecting. In just the same way as the notion of the hyper-real blurs the distinction in a society of signs between the real and the simulated, so the notion of hyper-connectivity blurs any distinction between the connected and the not connected or to put it another way, between the abstract and virtual and the concrete and face-to-face.

THE FACE-TO-FACE

The rapid spread of the Net and its seepage into everyday life raises fears of the effects of on-line existence where as people become electronically connected they also become disconnected from "real" life in the sense of structured relations of presence, raising fears of the dangerous erasure of the "real." On one hand, it could be said that virtual spaces have generated new public spheres and that new technological modes of experiencing and interaction are just as real and life enhancing as conversation. For Poster

(2001) the Net is not a tool but itself a form of community. Burbules (2000) argues that the Net needs to be understood as a virtual global community, a primary medium for the transmission of communication, information, culture, goods and services around the world. He sees the Net as a metacommunity, an overarching congregation of communities and a set of conditions that make communities possible, a space in which communities happen. The argument here then is that it is therefore a mistake to dismiss technology per se as dehumanizing or life negating and to only valorize non-technological activities and interaction as authentic and "human" (http://www.gseis.ucla.edu/people/pagre/tno/january-1995.html).

Others however query whether electronic connectivity can ever replace face-to-face interaction in meaningful communication. This is part of a wider debate where even if there really are communities in virtual spaces, these can be criticized as being fleeting and anonymous. It is argued that it is precisely this form of abstraction of electronic connectivity, the lack of the face-to-face that will always limit its potential in any educational space. However, it should be noted that these arguments are ones to do with democracy, participation and effective learning. Whilst they raise a number of legitimate concerns about the impact of ICTs they do not have quite the same resonance in considering the form taken by intellectual practices of knowledge production and the place of the face-to-face in these practices.

In broad terms, the issue of the virtual as against the face-to-face does not involve purely technological considerations but also ones of deep-rooted cultural understandings. As Poster (2001) points out, the Net is seen as eroding sociality and destabilizing community by undermining the face-to-face. The valorization of face-to-face communication is after all a cultural artifact, an example of what Derrida calls the metaphysics of presence, the humanistic tradition of the West that privileges the present. We have noted earlier Plato's criticism of writing in favour of the dialogue. This is the origin of the valorization of the face-to-face. Humanist shaped culture configures anything that is not face-to-face as a threat to an essential human condition.

But equally it could be argued that notions of the face-to-face as privileged modes of interaction are themselves practices constituted within particular discourses and cultures. Burbules (2000, p. 346) makes the point that it is "a cultural myth to imagine the more immediate interactions are always the most honest, open and intimate ones." Humanistic cultural understandings, although powerful, are historically contingent. In this context it is also worth recognizing that it is the growth of significance of the virtual that has highlighted the value of the face-to-face. As Poster (1995, p. 89) points out, the new kinds of interactivity that develop in cyberspace cannot be adequately specified by this binary and to think this way only serves "to obscure the manner of the historical construction of forms of community." He goes on to argue that what makes a community vital to its members is "their treatment of communications as meaningful and important" (Poster, 1995, p. 90). This is exactly what virtual communication signifies to its participants where the absence of the face-to-face is not regarded as a deficit.

Considering this in relation to intellectual practice and referring back to the earlier argument about writing as technological mediation, it is this form of mediation that characterizes the abstracted space where "standing back" is made possible. Abstraction does not require the presence of embodied others. On the contrary, it requires that the absence of the face-to-face is not seen as a deficit but as an advantage. An example is the printed text, a prime instance of writing as technological mediation whose effect is to provide access to knowledge without the need for face-to-face interaction with its producer. Of course, face-to-face interaction is an important component of knowledge production but my argument here is that it is not a *necessary* condition. The peer reviewing of journal articles and conference papers where the identity of the reviewer is unknown to the author is more typical and exemplary in this respect—there is an Other but that Other is virtual. It could be argued against this that collegiality as a version of the face-to-face has been a significant dimension in the intellectual practices of knowledge production and dissemination in the university and that collegiality is therefore a mark of the necessity of the face-to-face. Hinkson (2002) however points out that, although collegiality is a necessary feature of the connectivity that makes abstract interchange possible, it is not a necessary feature of the abstracted form of intellectual practice. It is simply a historically contingent way of organizing academic life and does not attenuate the abstraction of that form of life.

This debate over the place of the face-to-face is a valid one but I do find strangeness about it. What seems to be missed by the proponents of the face-to-face is that the immediacy of on-line connection suggests another form of the face-to-face. I would argue that one of the main characteristics of the face-to-face is immediacy and the Net provides immediacy albeit in a virtual way (http://www.emunix.emich.edu/~krause/Diss/). The implication I draw from this is that the face-to-face needs to be reconceptualized, particularly in terms of the notion of the necessity of *physical* and *ocular* presence. Equally, the notion of community needs to be reframed to take account of the fact that a community can no longer be defined simply in terms of geographical proximity or shared physical space.

FEAR AND LOATHING IN ACADEMIA

Cyberspace refers to the network of sites that can be accessed over the Net and the information available on that network. It is a spatialized metaphor that maps virtual relations on a global scale (http://www.ascusc.org/jcmc/vol1/issue2/aycock.html). It is a space without place where information in the form of data becomes accessible from almost anywhere on the globe. Accessible because of the global network of connectivity, but also because one doesn't have to be an expert to become part of that connectivity. Once connected, a variety of activities are possible and it is these activities that constitute the network of relationships that *is cyberspace.*

Cyberspace is also a term in a performative discourse that encapsulates a radical change in the way knowledge is understood and valorized:

- Electronic information is equated with knowledge
- Knowledge becomes globally transportable
- Knowledge becomes cast as intellectual property that can be bought and sold
- Knowledge is seen as essential to innovation
- Knowledge production does not take place only in universities.

With ICTs, the space of abstraction whose form, as we have seen, can be articulated as one of hyper-connectivity is different in kind rather than just in degree to the form of connectivity prior to the impact of ICTs. My purpose here is to highlight the fact that academics are not entirely sure about how to deal with cyberspace. In this sense there is a difference in terms of what technology has enabled. I'm not talking here about how to use e-mail or to surf the Net but about dealing with the dimensions of change noted above. The Net for example has made information more abundant and more accessible. The very speed and proliferation of information flows that is challenging definitions of what constitutes knowledge (Burbules, 2000), in the process questioning whether a distinction can and should be made between information and knowledge. I would argue that there is no definitive answer to this since these terms are embedded in competing discourses whose function is performative rather than truth seeking. To argue that information is not knowledge is not so much to make a statement about the nature of knowledge but to position oneself within an oppositional discourse resisting the encroachments of digitization and commercialization. As the Net enables knowledge to be globally generated and distributed, academics are losing their power to define what constitutes knowledge and their dominance over the production and dissemination of knowledge. What constitutes knowledge is now not bound and thus defined by the disciplinary canons set by the university. Academics are losing their monopoly over expert knowledge. Disciplinary knowledge itself is found in abundance on the Net but so too all kinds of other knowledge flourish in that virtual space. Any group with access to the Net and with minimal computer literacy can develop their own (virtual) institutions and engage in (virtual) publication of their knowledge. Anyone with minimal technical expertise and at minimal cost can create their own website and put whatever they want on it and someone somewhere is going to access it and find it useful. And for some, this may be "mere" information but for others it will be seen as knowledge. This trend is propelled by the very economy of the Net. This is an economy where rather than more being less the opposite is the case. More is no more costly than less with the value of the Net increasing with every increase in users. There is a built-in dynamic for exponential growth, for infinite connectivity.

The Net therefore enables more information to be distributed more widely. It also enables different kinds of knowledge to flourish as well as providing tools for new approaches to knowledge (http://www.press.umich.edu/jep/06-01/mack.html). Gibbons et al. (1994) refer to the increasing significance of Mode 2 knowledge production—socially distributed knowledge produced in a variety of sites. This is a type of knowledge produced within communities of practice that are not necessarily located within the university. They characterize this mode of knowledge production as trans-disciplinary and trans-institutional, meaning that it not only cuts across disciplines but also uses them eclectically and without respect for traditional forms and boundaries. By trans-institutional they refer to the fact that "those outside the university can become active agents in the definition and solution of problems as well as in the evaluation of performance" (Gibbons et al., 1994, p. 7). As we have seen in the case of the techno sciences, there is no longer a linear process of discovery to application. Discovery, once the exclusive province of the university, can now occur anywhere and no longer needs the university's mark of legitimacy.

The significance of these changes lies in the kind of social space they create. As I have already implied, the contemporary moment is one where new ways of thinking about different kinds of social space are required. The university cannot remain indifferent to this need. The problem is that what the university has traditionally taken for granted can function both as a bank of accumulated wisdom, a wisdom that can be harnessed to the cause of change, or as mindless resistance to change (http://dlis.gseis.ucla.edu/people/pagre/newcastle.html). Either way, change in the sense of reframing, is difficult and complex. Reframing can have a variety of content and shape, creating different spaces according to different modes of interaction. At the same time, this reframing needs to be located within the changes wrought by ICTs since the very need for reframing comes about because of the transformation and creation of new spatialities and relationships (http://www.lmappingcyberspace.com).

With the university and its particular form of intellectual practice, the reframing is to do with boundaries and hierarchies. In broad terms we can say that spatial boundaries both vertically and horizontally are now much more porous and traversable with greater ease. There are two types of boundaries that are changing in this way—the boundaries between disciplines and the boundaries between disciplinary and other types of knowledge. Even within the university disciplines are less clearly demarcated and have lost much of their organizational and epistemological potency. Outside of the university, performativity has rendered the traditional disciplines less of an anchoring and reference point for all knowledge production. However, this is not just a matter of the Net stimulating the breakdown of the boundedness of disciplinary knowledge and disciplinary definitions of what constitutes knowledge. There is also an issue that is to do with what the Net stimulates in terms of the form of writing. Here we have to note the significance and impact of hypertext. Burbules (1997) puts it this way: "the conventions of writing… assume a fundamentally linear and hierarchical organization of information with passage following passage in a sequence governed by (a) the relative importance, formalized in the discipline of the outline and (b) the narrative structure of argument formalized in the discipline of the syllogism…" (http://faculty.ed.uiuc.edu/burbules/ncb/papers/rhetorics.html). The outline and the syllogism have been the predominant form of writing in the West since the scrolled parchment. They have defined the space of writing characterizing intellectual practices in a time where disciplines were dominant. Hypertext marks a radical difference. The form of writing appropriate to hypertext is, argues Burbules, the bricolage and the juxtaposition. These new forms of writing stimulate a different pragmatics of reading, what Burbules terms "hyper-reading." The growth of this practice of reading is likely to make its contribution to the increasing porosity of boundaries.

Intellectual practices have always been hierarchical. In universities there are different interlocking hierarchies—organizational, academic and epistemological. There are hierarchies between and within disciplines. All of these are now in the process of rapid and bewildering change. The new breed of managers increasingly dominate organizational hierarchies. "In reinventing research as a system of money, measures, targets and comparisons, amidst a competitive economy and an enterprise culture, research management sets out to change the way that researchers think about research and how they see themselves as researchers" (Marginson & Considine, 2000, p. 164). The techno sciences are fast taking over the dominant position within the university traditionally occupied by the natural sciences and the humanities. The commercialization of knowledge and the move from the interpretive to the instrumental is subverting the traditional academic logic of enquiry and discovery. "Tertiary education policies in all countries [have] moved towards... policies that emphasize academic capitalism at the expense of basic or fundamental research, towards curricula policy that concentrated moneys in science and technology and fields close to the market" (Slaughter & Leslie, 1997, p. 55). The role of universities in selecting elites is breaking down with the massification and vocationalization of higher education, and the role of providing a rite of passage is beginning to disappear as off campus and online learning becomes more widespread (http://www.futuresproject.org).

There is another aspect of hierarchy that needs to be noted. This is to do with the performativity of the Net. A major characteristic of ICTs is that in principle they compress time, (although because of bandwidth problems not always in practice) and communication becomes virtually instantaneous. This removal of a time constraint has of course enormous advantages for the dissemination of knowledge texts but at the same time it undermines the authority of the writer. Poster (2001, p. 92) puts it this way: "embodied in computer files, digital texts subsist in space only at the whim of the reader. The author of digital texts loses the assurance of their spatial continuity. Pages of digital text have the stability of liquid. They may be altered in their material arrangements of traces as they are read." There are two points to be made here. The first is that texts on the Net are both time-less and space-less. This is another way of saying that they are both ephemeral and never closed. For academics, both these characteristics are difficult to accept, accustomed as they are in their intellectual practice in embodying their knowledge production in printed texts (books, journal articles, published conference proceedings), which have solidity, permanence, continuity and a closure. The printed texts of academics are "objects in space... for long periods, they are enough the same to yield themselves to different readers with the exact display of traces" (Poster, 2001, p. 92). With digital texts this is no longer the case. This leads on to the second point, which is to do with the author and with author-ity. As Bolter (1991) notes, the Net works against the fixity of texts. It functions to transfer authorship and thereby authority from the writer to the reader, a process of decentring where the construction of meaning falls onto readers who can manoeuvre in whatever sequence or fashion they choose (Ess, 1996) (http://www.hanover.edu/philos/ejournal/archive/v6n3/ess/ess.html). Even if academics work increasingly online with digital texts therefore, without spatial continuity and stability their texts are no longer as authoritative, or at least the source of authority changes since now it is what the online virtual reader makes of them that determines their significance. This relates to the argument made earlier about the new practices of reading that flow from hyper textual practices of writing centred on bricolage and juxtapositon.

The problem now however is that, regardless of the view we might have of hierarchies, they are nonetheless a structuring form and, along with boundaries, constitute the cement of the space within which academics have located themselves—through which they get their bearings, a sense of direction and a mapping in relation to their fellow academics and the world outside academia. Identity is constructed through practices of boundary marking and hierarchization, practices that divide the self from the other and therefore enable a self to be defined. In turn forms of identity and changes in these forms can reinforce or unsettle boundaries and hierarchies (http://bates.cstudies.ubc.ca/carnegie/Carnegie.html). These changes are shaped by the impact of ICTs. Academic identities are to a significant extent now being constructed through a mode of communication, which as we have seen goes beyond the instrumental. In an increasingly globalized and wired (connected) world and with all the consequences that state has for the production, dissemination and valuation of knowledge, academics find that their identity is becoming confused and uncertain. The work of reframing they need to do to locate them in a reconstituted space is still in process. A new space is being shaped where identity is being rewritten. Whilst the form of intellectual practice still consists of an apparently familiar "standing back," the nature of this is more that of hyper-connectivity with related features that mirror the impact of ICTs and the onset of the information society and knowledge economy. With this has come the porosity of boundaries and the breakdown of existing hierarchies and their reconstitution in different forms. As Poster (2001) argues the Net becomes an alien Other, both fascinating and repellent. Hence the fear and loathing which is so characteristic of the contemporary university.

VIRTUALLY ... THE END

The form of intellectual practice is still an abstracted space. However the proliferation of ICTs means that the technological mediation that now characterizes intellectual practice is that of hyper-connectivity and here my argument has been that this is not simply a difference of degree but more a significant difference in kind. As I have noted earlier, hyper-connectivity refers to space where there is always already connection such that it is inconceivable that a space could exist where this is not the case. The structure of this always already connected space is that of the network where "network" refers both to the complex patterning of interactions and positioning that now takes place on a global scale and to the Net itself, which makes this complex network possible and is itself an effect of that patterning. It is in this sense that the evolving abstracted space of intellectual practice is radically different. It is therefore not so much the standing back that uniquely defines this space, although that is still present, but the hyper-connectivity, connectivity stimulated by new practices of writing and reading enabled by the Net.

Hyper-connectivity has implications for the face-to-face and in particular for the diminished importance of relations of presence. Many regard this as a disturbing trend. This is hardly surprising once the power of cultural and pedagogical valorizations of such relations is taken into account. This disturbance has been particularly acute in the university where the face-to-face and the collegial have now become central elements in the resistance to performativity and all that goes with it. To me, this itself is an indicator that the abstracted space of hyper-connectivity is radically different to previous forms of abstracted space that characterized the intellectual practices of the university since if this were not the case it would not have led to such disturbance nor occasioned such resistance. For there to be resistance there must be something to resist.

The conclusion I draw from this is that the university continues to be a site of contestation with the form of intellectual practice the stake in the contest. In one sense there is nothing particularly new about this since the university has historically been such a site. Furthermore, the stake in the contestation has been the form that practice should take and the appropriate place of technological mediation. However, in another sense, ICTs, the Net and virtuality introduce new and destabilizing elements that are present not just in the university but also in the social form generally. Ultimately the conflict is over the space to be occupied and how space is to be defined, an issue that becomes acute when the conflict inevitably involves coming to terms with the spaceless space that virtuality introduces.

I would argue then in conclusion that it is necessary to find a balance between the abstraction of virtuality and the presence of the face-to-face. How this is to be done is beyond the scope of this chapter but I would venture to suggest that a possible way forward might involve rethinking the notion of the face-to-face and recognizing that, within the abstracted space of the virtual, new forms of relationality, new forms of immediacy, are both possible and necessary.

REFERENCES

Bolter, J, D. (1991). *Writing space: The computer, hypertext, and the history of writing.* Hillsdale, NJ: Laurence Erlbaum Associates.

Borgmann, A. (1992). *Crossing the postmodern divide.* Chicago: University of Chicago Press.

Bowen, W. (2000). *At a slight angle to the universe: The university in a digitalized, commercialised age.* Princeton: Princeton University Press.

Burbules, N. (1997). Rhetorics of the web: Hyperreading and critical literacy. In I. Snyder (Ed.), *Page to screen: Taking literacy into the electronic era.* NSW: Allen and Unwin.

Burbules, N. (2000). Does the internet constitute a global educational community? In N.Burbules & C.Torres (Eds.), *Globalization and education: Critical perspectives.* London: Routledge.

Carnoy, M. (2000). Globalization and educational reform. In N.P.Stromquist & K. Monkman (Eds.), *Globalisation and education: Integration and contestation across cultures.* Oxford: Rowman and Littlefield.

Castells, M. (1996). *The rise of the network society.* Oxford: Blackwell.

Edwards, R., & Usher, R. (1999). A society of signs? Mediating a learning society. *British Journal of Educational Studies, 47*(3), 261–274.

Ess, C. (1996). Modernity and postmodernity in 'hypertext notes'. *Ejournal, 6*(3).

Gibbons, M., Limoges, C., Nowotny, H., Schwartzman, S., Scott, P., & Trow, M. (1994). *The new production of knowledge: The dynamics of science and research in contemporary societies.* London: Sage.

Hasan, R. (2001). Net results: Knowledge, information and learning on the internet. *Journal of Educational Enquiry, 2*(2), 44–57.

Hinkson, J. (2002). Perspectives on the crisis of the university. In S. Cooper, J. Hinkson, & G. Sharp (Eds.), *Scholars and entrepreneurs: The universities in crisis.* pp. 260–262. Melbourne: Arena Publications.

Jordan, T. (1999). *Cyberpower: The culture and politics of cyberspace and the internet.* London: Routledge.

Lyotard, J-F. (1984). *The postmodern condition: A report on knowledge.* Manchester: Manchester University Press.

Marginson, S., & Considine, M. (2000). *The enterprise university.* Cambridge: Cambridge University Press.

Poster, M. (1995). Postmodern virtualities. In M. Featherstone & R. Burrows (Eds.), *Cyberspace, cyberbodies, cyberpunk: Cultures of technological embodiment.* pp. 89–90. London: Sage.

Poster, M. (2001). *What's the matter with the internet?* Minneapolis: University of Minnesota Press.

Sharp, G. (2002). The idea of the intellectual and after. In S. Cooper, J. Hinkson, & G. Sharp (Eds.), *Scholars and entrepreneurs: The universities in crisis.* Melbourne: Arena Publications.

Slaughter, S., & Leslie, L. (1997). *Academic capitalism: Politics, policies and the entrepreneurial university.* Baltimore: Johns Hopkins University Press.

www.ingramcontent.com/pod-product-compliance
Lightning Source LLC
Chambersburg PA
CBHW061608180926
51615CB00006B/21